About Island Press

Island Press, a nonprofit organization, publishes, markets, and distributes the most advanced thinking on the conservation of our natural resources—books about soil, land, water, forests, wildlife, and hazardous and toxic wastes. These books are practical tools used by public officials, business and industry leaders, natural resource managers, and concerned citizens working to solve both local and global resource problems.

Founded in 1978, Island Press reorganized in 1984 to meet the increasing demand for substantive books on all resource-related issues. Island Press publishes and distributes under its own imprint and offers these services to other nonprofit organizations.

Support for Island Press is provided by Apple Computer, Inc., Geraldine R. Dodge Foundation, The Energy Foundation, The Charles Engelhard Foundation, The Ford Foundation, Glen Eagles Foundation, The George Gund Foundation, William and Flora Hewlett Foundation, The Joyce Foundation, The John D. and Catherine T. MacArthur Foundation, The Andrew W. Mellon Foundation, The Joyce Mertz-Gilmore Foundation, The New-Land Foundation, The J. N. Pew, Jr. Charitable Trust, Alida Rockefeller, The Rockefeller Brothers Fund, The Rockefeller Foundation, The Florence and John Schumann Foundation, The Tides Foundation, and individual donors.

Learning to Listen to the Land

Edited by Bill Willers

Learning to Listen to the Land

Foreword by David Brower

ISLAND PRESS
Washington, DC □ *Covelo, California*

For Karl H. Willers and
Virginia Brown Willers

Text design by David Bullen

Library of Congress Cataloging-in-Publication Data

Learning to listen to the land / [selected by] Bill Willers.

p. cm.
Includes index.
ISBN 1-55963-121-X (cloth).—ISBN 1-55963-120-1 (paper)
1. Human ecology. 2. Human ecology—Moral and ethical aspects. 3. Environmental protection—Moral and ethical aspects.
I. Willers. W. B., 1938–
GF49.L43 1991
179′.1—dc20 91-21890
CIP

Printed on recycled, acid-free paper

Manufactured in the United States of America
10 9 8 7 6 5 4

Where were you when I laid the foundation of the earth?

Job 38:4

Contents

Foreword

If Wallace Stegner and friends want us to learn to *listen* to the land, and they do, I'll go along, adding that we should listen eloquently. More than that, we should learn to *see* the land—to look at the Earth as if we had never seen it before. We can also put our other senses to work. With *touch* we can feel its textures: the warmth, the coolness; what is sharp and hurts, what is smooth and soothes. With our sense of *smell* we can try to get away from artifacts and learn from nature's aromas, and what they mean for *taste*.

Heed another sense, *intuition*. Through its eons of learning, intuition can outwit any computer. It can absorb enormously complex data in no time at all and instantly flash back to you its well-informed appraisal: OK, Error, or Not Found. There is still another sense, *reason*. This sense is probably incredibly junior to intuition, but few people think so yet. It would help if everyone would look hard at the progress reason has wrought.

A tough reappraisal of progress is overdue. Unquestioned growth is not part of it. As I put it twenty-five years ago, we ought not to be blindly against progress, but against blind progress, and to try to distinguish one from the other, which few people do.

It is all right to be sad about the predicament of industrialized humanity, but not to linger long in sadness. It is all right to despair, but not to remain despairing. It is all right to be wrong, but not to stay wrong, as former Sierra Club Southwest Representative Jeff Ingram said in 1968. And it is all right to ask for help.

You will find that help right here.

Learning to Listen to the Land is a symphony of voices that can bring enormous help, if heeded, to all your senses and to civilization's future. Indeed, no important decisions should be made by anyone until he or she has read this book.

The person who first comes to mind who needs to read and understand this book is Environmental President George Bush. Next would be the new—although not in their thinking—majority of the Supreme Court. Concurrently, it should be required reading for all other branches of this and other governments, for CEOs and directors and MBAs of all corporations, for faculties and administrators of all universities, for leaders of all religions, for writers and editors and publishers of all prose, poetry, pulp, and hype, as well as for all who have been influenced by any of the above. Have I left anyone out?

To print that many copies of this book, I realize, would place the world's remaining forests in unprecedented peril, which leaves three possible solutions: lend the book with a two-day limit; broadcast it on National Public Radio and Voice of America; or make a miniseries of it for television with undivided attention required of all audiences, worldwide.

Of course there are some things wrong with the book. It almost totally lacks the condiment of humor, and if reform can't be fun part of the time it isn't going to happen. A chapter or two are too high on the abstraction ladder to get through to people, like me, who need them most. And it has been left to me to insist that all readers memorize the chapters by Wallace Stegner, Herman Daly, and Carolyn Merchant, then put their recommendations into effect.

Memorizing shouldn't be too hard. Marion Edy, who made the League of Conservation Voters become an important, new institution, committed *The Ancient Mariner* to memory—and recited it to me to prove it. I myself remembered that the poem gave the albatross a bad name. *Learning to Listen to the Land* tells us this metaphor plagues us all worse than ever. I memorized only eleven lines—including "Water, water everywhere,/Nor any drop to drink"—without realizing how imminent this situation would be on the only planet that we know has water on board.

Going through the chapters, I made special note of whatever had anything to do with getting global restoration under way, realizing with the United Nations Environment Program's Noel Brown that "It is healing time on Earth" (his title for a speech; I have borrowed it ever since) and time to replace Operation Desert Storm with Operation Desert Flower. Wendell Berry, you will read herein, doesn't think this healing is possible. He is right about so many things. How can he be so wrong on restoration?

Carolyn Merchant's chapter, with admirable lucidity, lets you see how wrong he is. All she fails to explain is how huge, capital-intensive, multiracial, multinational, multibelief, multisex, absorbing, exciting, and enjoyable the opportunity for diverse restoration teams to work together can and must be, and how soon the industrial nations owe it to the rest of the world to get it going with no further delay. It can be a splendid substitute for war.

To expedite restoration, let it be profitable. The profit can be enjoyed now and would be shared as well by the seventh generation and beyond: those sentient beings yet to arrive on Earth, beings which indigenous people, in their reverence for ethics, manage to keep in mind while the rest of us cash in on their options.

If capitalism can make the Green Century of Global Restoration work, then all may be forgiven. The same goes for any other species of government. If none of them can, it's high time for the first real paradigm shift since we discovered that the world was round, not flat. Now is the time for all men and women to admit that the Earth is not only round, but is also limited, and not here just for us.

If existing forms of government can't make that paradigm shift—and they show little promise of being able to—it is time for the Green Shift.

And here is its handbook!

—*David R. Brower*
Chairman, Earth Island Institute

Learning to Listen to the Land

Introduction

In recent years, as it has become increasingly apparent that human societies are operating in ways that are destructive to the planet, there has been an outpouring of fine writing built upon the legacy of such writers as Henry David Thoreau, John Muir, and Aldo Leopold. Much of this new work has come from scientists and others in academic areas traditionally associated with an objective outlook.

In addition, we are seeing a shift in preoccupation from the narrow concerns of individual specialties toward interconnectedness. As this holistic viewpoint grows, so does recognition that the purely rational attitude that for so long has dominated our civilization has excluded spiritual considerations in our understanding of the earth-human relationship—an exclusion that may be largely responsible for the ecological dilemma in which we find ourselves. As Paul Ehrlich wrote in 1986, "Scientific analysis points, curiously, toward the need for a quasi-religious transformation of contemporary cultures."

Any selection of the sort presented in this book will reflect the selector's worldview. My own perspective derives from the fact that I am a biologist who, having spent years witnessing planetary deterioration and having sought reasons for that deterioration, has seen the undeniable evidence that fundamental characteristics of our civilization are at fault. We have, over time, wittingly or not, generated a society characterized by greed, exploitation, and a presumption of perpetual growth. In the process, we have become psychologically detached from Nature and have transferred our affections, awe, and respect from the natural world to the material goods that we produce. So bewitched are we by our technical prowess, and by the creature comfort that stems from it, that belief in its ability to overcome all problems has become a kind of religion. When adherents of this religion speak of the "real world," they are not talking about biology or natural cycles or anything concerning a living planet that has been eons in the formation. Rather, they are referring to the contemporary business atmosphere. It is a monument to human arrogance that an economic system—a mere human construct of relatively recent design—has become what is most real in life.

The essays contained in this book were selected in order to paint in words a portrait of the planet that I have not found elsewhere in the literature. It is a picture that treats biological diversity and natural processes but one that also is grounded in deep, aesthetic appreciation as well as in the practicality of survival considerations. It reveals a perspective that I want my students to have—students were in mind as I made selections—but it's a perspective, I'm convinced, that a majority of people must cultivate before planetary healing can begin. I want readers to know that The Garden, beleaguered though it may be, never really left us. It is we who drifted away and we who, through a fog of separation and materialism, lost our ability to see clearly.

The anthology is divided into three parts. The first presents a view of the living earth and some of its traits and processes. The central focus in this part is on the diversity of life forms and on the relatedness and interdependence of those forms—concepts contained within the word "biodiversity." Throughout, one finds a strong sense that there is inherent rightness in the rich variety that is a gift of the ages.

The second part deals with the impact on the earth of the human species, with comments by writers from a variety of backgrounds. Regardless of background, though, concern and even anger come through. And because the writers are reflections of the society in which they live, they indicate growing concerns within the general public.

Whereas the second part paints a dismal picture, the third is hopeful, because it reveals that within society there are many people struggling for constructive change. Their commonality is an obvious sense of connection with the earth, and their approaches and solutions, therefore, are in keeping with earth's processes, rather than counter to them. What these people need are allies. What they know must be known by many more. Only when their knowledge and views have permeated society and become the accepted norm will needed change become reality.

—*Bill Willers*

PART ONE

Nature and Biodiversity

Accuse not nature, she hath done her part; do thou but thine.

—John Milton

Paradise Lost

Becoming familiar with the Gaia Hypothesis and then ruminating about it for some months foster an attitude toward the planet that is at odds with the prevailing attitude. The notion of Earth as a living entity, rather than as a big spherical rock with living creatures on *it, is a kind of heresy in a society geared, as is ours, to exploitation.*

But the notion is seductive, and once planted it flourishes. There is an inherent rightness about viewing Earth, hanging in the void, vibrantly ruddy and blue, wreathed in cottony swirls, as alive. It's as though the idea wasn't really new for us—that it was there all along, only temporarily suppressed, needing but a faint nudge to blossom again in the mind. The very thought elicits a sense of unity; holism becomes inescapable.

A common misconception about Gaia, though, is that, in the manner of a good mother, it must have a special fondness for us humans, and that it exists primarily to nurture our species. This may be a normal assumption for beings who have become as divorced from the natural world as we have become. But to hold to such a view is to avoid the central point that Gaia's processes are directed toward maintaining the well-being of Gaia, and there is every indication that, were it necessary in order to maintain its own good health, Gaia would dispatch us as dispassionately as we might wipe out weeds in a garden.

The realization that from the standpoint of Gaia we are co-equal with other resident species is a step that we must take before it is possible to see clearly that we are parts of something rather than masters of something. The realization is both humbling and comforting. We enjoy, we use, we belong; we are not, however, at the helm.

Having taken such a mental journey, we are able to apprehend wilderness for what it really is—not a collection of parks for our vacation enjoyment, but intact ecosystems that, like Gaia itself, are discrete entities having inherent integrity. Intuition will then tell us that all constituent parts, from microbes to redwoods, from worms to top-level predators, from groundwaters to mountaintops, have rights to exist by virtue of the fact that Gaia's processes have caused them to come into being.

A sketch by Sir James Lovelock of his concept of Gaia begins the part. Thereafter, pieces by E. O. Wilson, Reed Noss, and Chris Maser deal with biological diversity on the planet. While Wilson takes a global view, Maser concentrates on the concept of ancient forest, and Noss focuses on ill-conceived policy within our country that, in favoring diversity of "weed species" in local pockets, actually diminishes diversity overall. Norman Myers then writes about the implications of impending mass extinctions. And Joan Bird's essay brings together information about the fragmentation of ecosystems, the resulting isolation of "island populations," and the dangers faced by those populations that have become genetically isolated.

The last two essays deal with human attitudes. David Ehrenfeld's "The Conservation Dilemma" is from his book The Arrogance of Humanism, *this latter title neatly describing his thesis. Humanists, he contends, are not willing to save any "fragment of Nature" that lacks usefulness to humankind. Wallace Stegner's "The Gift of Wilderness" is the best of possible pieces to end this part. One of the finest prose stylists ever to write about Nature, he makes a point that, in many ways, is the point of the anthology. "We need," he maintains, "to learn to listen to the land."*

—BW

The Earth as a Living Organism

James E. Lovelock

The idea that the Earth is alive may be as old as humankind. The ancient Greeks gave her the powerful name Gaia and looked on her as a goddess. Before the nineteenth century even scientists were comfortable with the notion of a living Earth. According to the historian D. B. McIntyre (1963), James Hutton, often known as the father of geology, said in a lecture before the Royal Society of Edinburgh in the 1790s that he thought of the Earth as a superorganism and that its proper study would be by physiology. Hutton went on to make the analogy between the circulation of the blood, discovered by Harvey, and the circulation of the nutrient elements of the Earth and of the way that sunlight distills water from the oceans so that it may later fall as rain and so refresh the earth.

This wholesome view of our planet did not persist into the next century. Science was developing rapidly and soon fragmented into a collection of nearly independent professions. It became the province of the expert, and there was little good to be said about

Biodiversity (Washington, D.C.: National Academy Press, 1988).

interdisciplinary thinking. Such introspection was inescapable. There was so much information to be gathered and sorted. To understand the world was a task as difficult as that of assembling a planet-size jigsaw puzzle. It was all too easy to lose sight of the picture in the searching and sorting of the pieces.

When we saw a few years ago those first pictures of the Earth from space, we had a glimpse of what it was that we were trying to model. That vision of stunning beauty; that dappled white and blue sphere stirred us all, no matter that by now it is just a visual cliché. The sense of reality comes from matching our personal mental image of the world with that we perceive by our senses. That is why the astronaut's view of the Earth was so disturbing. It showed us just how far from reality we had strayed.

The Earth was also seen from space by the more discerning eye of instruments, and it was this view that confirmed James Hutton's vision of a living planet. When seen in infrared light, the Earth is a strange and wonderful anomaly among the planets of the solar system. Our atmosphere, the air we breathe, was revealed to be outrageously out of equilibrium in a chemical sense. It is like the mixture of gases that enters the intake manifold of an internal combustion engine, i.e., hydrocarbons and oxygen mixed, whereas our dead partners Mars and Venus have atmospheres like gases exhausted by combustion.

The unorthodox composition of the atmosphere radiates so strong a signal in the infrared range that it could be recognized by a spacecraft far outside the solar system. The information it carries is prima facie evidence for the presence of life. But more than this, if the Earth's unstable atmosphere was seen to persist and was not just a chance event, then it meant that the planet was alive—at least to the extent that it shared with other living organisms that wonderful property, homeostasis, the capacity to control its chemical composition and keep cool when the environment outside is changing.

When on the basis of this evidence, I reanimated the view that we were standing on a superorganism rather than just a ball of rock (Lovelock, 1972; 1979), it was not well received. Most scientists either ignored it or criticized it on the grounds that it was not needed to explain the facts of the Earth. As the geologist H. D. Holland

(1984, p. 539) put it, "We live on an Earth that is the best of all possible worlds only for those who are well adapted to its current state." The biologist Ford Doolittle (1981) said that keeping the Earth at a constant state favorable for life would require foresight and planning and that no such state could evolve by natural selection. In brief, scientists said, the idea was teleological and untestable. Two scientists, however, thought otherwise; one was the eminent biologist Lynn Margulis and the other the geochemist Lars Sillen. Lynn Margulis was my first collaborator (Margulis and Lovelock, 1974). Lars Sillen died before there was an opportunity. It was the novelist William Golding (personal communication, 1970), who suggested using the powerful name Gaia for the hypothesis that supposed the Earth to be alive.

In the past ten years these criticisms have been answered—partly from new evidence and partly from the insight provided by a simple mathematical model called Daisy world. In this model, the competitive growth of light- and dark-colored plants on an imaginary planet is shown to keep the planetary climate constant and comfortable in the face of a large change in heat output of the planet's star. This model is powerfully homeostatic and can resist large perturbations not only of solar output but also of plant population. It behaves like a living organism, but no foresight or planning is needed for its operation.

Scientific theories are judged not so much by whether they are right or wrong as by the value of their predictions. Gaia theory has already proved so fruitful in this way that by now it would hardly matter if it were wrong. One example, taken from many such predictions, was the suggestion (Lovelock, 1972) that the compound dimethyl sulfide would be synthesized by marine organisms on a large scale to serve as the natural carrier of sulfur from the ocean to the land. It was known at the time that some elements essential for life, like sulfur, were abundant in the oceans but depleted on the land surfaces. According to Gaia theory, a natural carrier was needed and dimethyl sulfide was predicted. We now know that this compound is indeed the natural carrier of sulfur, but at the time the prediction was made, it would have been contrary to conventional wisdom to seek so unusual a compound in the air and the sea. It is unlikely that its presence would have been sought but for the stimulus of Gaia theory.

Gaia theory sees the biota and the rocks, the air, and the oceans as existing as a tightly coupled entity. Its evolution is a single process and not several separate processes studied in different buildings of universities.

It has a profound significance for biology. It affects even Darwin's great vision, for it may no longer be sufficient to say that organisms that leave the most progeny will succeed. It will be necessary to add the proviso that they can do so only so long as they do not adversely affect the environment.

Gaia theory also enlarges theoretical ecology. By taking the species and the environment together, something no theoretical ecologist has done, the classic mathematical instability of population biology models is cured.

For the first time, we have from these new, these geophysiological models a theoretical justification for diversity, for the Rousseau richness of a humid tropical forest, for Darwin's tangled bank. These new ecological models demonstrate that as diversity increases so does stability and resilience. We can now rationalize the disgust we feel about excesses of agribusiness. We have at last a reason for our anger over the heedless deletion of species and an answer to those who say it is mere sentimentality.

No longer do we have to justify the existence of the humid tropical forests on the feeble grounds that they might carry plants with drugs that could cure human disease. Gaia theory forces us to see that they offer much more than this. Through their capacity to evapotranspire vast volumes of water vapor, they serve to keep the planet cool by wearing a sunshade of white reflecting clouds. Their replacement by cropland could precipitate a disaster that is global in scale.

A geophysiological system always begins with the action of an individual organism. If this action happens to be locally beneficial to the environment, then it can spread until eventually a global altruism results. Gaia always operates like this to achieve her altruism. There is no foresight or planning involved. The reverse is also true, and any species that affects the environment unfavorably is doomed, but life goes on.

Does this apply to humans now? Are we doomed to precipitate a change from the present comfortable state of the Earth to one almost certainly unfavorable for us but comfortable to the new biosphere of our successors? Because we are sentient there are alternatives, both good and bad. In some ways the worse fate in store for us is that of becoming conscripted as the physicians and nurses of a geriatric planet with the unending and unseemly task of forever seeking technologies to keep it fit for our kind of life—something that until recently we were freely given as a part of Gaia.

Gaia philosophy is not humanist. But being a grandfather with eight grandchildren I need to be optimistic. I see the world as a living organism of which we are a part; not the owner, nor the tenant, not even a passenger. To exploit such a world on the scale we do is as foolish as it would be to consider our brains supreme and the cells of other organs expendable. Would we mine our livers for nutrients for some short-term benefit?

Because we are city dwellers, we are obsessed with human problems. Even environmentalists seem more concerned about the loss of a year or so of life expectation through cancer than they are about the degradation of the natural world by deforestation or greenhouse gases—something that could cause the death of our grandchildren. We are so alienated from the world of nature that few of us can name the wild flowers and insects of our locality or notice the rapidity of their extinction.

Gaia works from an act of an individual organism that develops into global altruism. It involves action at a personal level. You well may ask, So what can I do? When seeking to act personally in favor of Gaia through moderation, I find it helpful to think of the three deadly Cs: combustion, cattle, and chain saws. There must be many others.

One thing you could do, and it is no more than an example, is to eat less beef. If you do this, and if the clinicians are right, then it could be for the personal benefit of your health; at the same time, it might reduce the pressures on the forests of the humid tropics.

To be selfish is human and natural. But if we chose to be selfish in the right way, then life can be rich yet still consistent with a world fit for our grandchildren as well as those of our partners in Gaia.

References

Doolittle, W. F. 1981. Is nature really motherly? CoEvol. Q. 29:58–63.

Holland, H. D. 1984. The Chemical Evolution of the Atmosphere and the Oceans. Princeton University Press, Princeton, N.J. 656 pp.

Lovelock, J. E. 1972. Gaia as seen through the atmosphere. Atmos. Environ. 6:579–580.

Lovelock, J. E. 1979. Gaia. A New Look at Life on Earth. Oxford University Press, Oxford. 157 pp.

McIntyre, D. B. 1963. James Hutton and the philosophy of geology. Pp. 1–11 in Claude C. Albritton, ed. The Fabric of Geology. Addison-Wesley, Reading, Mass.

Margulis, L., and J. E. Lovelock. 1974. Biological modulation of the Earth's atmosphere. Icarus 21:471–489.

The Current State of Biological Diversity

E. O. Wilson

Biological diversity must be treated more seriously as a global resource, to be indexed, used, and above all, preserved. Three circumstances conspire to give this matter an unprecedented urgency. First, exploding human populations are degrading the environment at an accelerating rate, especially in tropical countries. Second, science is discovering new uses for biological diversity in ways that can relieve both human suffering and environmental destruction. Third, much of the diversity is being irreversibly lost through extinction caused by the destruction of natural habitats, again especially in the tropics. Overall, we are locked into a race. We must hurry to acquire the knowledge on which a wise policy of conservation and development can be based for centuries to come.

To summarize the problem in this chapter, I review some current information on the magnitude of global diversity and the rate at which we are losing it. I concentrate on the tropical moist forests, because of all the major habitats, they are richest in species and because they are in greatest danger.

Biodiversity (Washington, D.C: National Academy Press, 1988)

THE AMOUNT OF BIOLOGICAL DIVERSITY

Many recently published sources, especially the multiauthor volume *Synopsis and Classification of Living Organisms*, indicate that about 1.4 million living species of all kinds of organisms have been described (Parker, 1982; see also the numerical breakdown according to major taxonomic category of the world insect fauna prepared by Arnett, 1985). Approximately 750,000 are insects, 41,000 are vertebrates, and 250,000 are plants (that is, vascular plants and bryophytes). The remainder consists of a complex array of invertebrates, fungi, algae, and microorganisms (see Table 1). Most systematists agree that this picture is still very incomplete except in a few well-studied groups such as the vertebrates and flowering plants. If insects, the most species-rich of all major groups, are included, I believe that the absolute number is likely to exceed 5 million. Recent intensive collections made by Terry L. Erwin and his associates in the canopy of the Peruvian Amazon rain forest have moved the plausible upper limit much higher. Previously unknown insects proved to be so numerous in these samples that when estimates of local diversity were extrapolated to include all rain forests in the world, a figure of 30 million species was obtained (Erwin, 1983). In an even earlier stage is research on the epiphytic plants, lichens, fungi, roundworms, mites, protozoans, bacteria, and other mostly small organisms that abound in the treetops. Other major habitats that remain poorly explored include the coral reefs, the floor of the deep sea, and the soil of tropical forests and savannas. Thus, remarkably, we do not know the true number of species on Earth, even to the nearest order of magnitude (Wilson, 1985a). My own guess, based on the described fauna and flora and many discussions with entomologists and other specialists, is that the absolute number falls somewhere between 5 and 30 million.

A brief word is needed on the meaning of species as a category of classification. In modern biology, species are regarded conceptually as a population or series of populations within which free gene flow occurs under natural conditions. This means that all the normal, physiologically competent individuals at a given time are capable of breeding with all the other individuals of the opposite sex belonging

Table 1 *Numbers of Described Species of Living Organisms*[a]

Kingdom and Major Subdivision	Common Name	No. of Described Species	Totals
Virus			
	Viruses	1,000 (order of magnitude only)	1,000
Monera			
Bacteria	Bacteria	3,000	
Myxoplasma	Bacteria	60	
Cyanophycota	Blue-green algae	1,700	4,760
Fungi			
Zygomycota	Zygomycete fungi	665	
Ascomycota (including 18,000 lichen fungi)	Cup fungi	28,650	
Basidiomycota	Basidiomycete fungi	16,000	
Oomycota	Water molds	580	
Chytridiomycota	Chytrids	575	
Acrasiomycota	Cellular slime molds	13	
Myxomycota	Plasmodial slime molds	500	46,983
Algae			
Chlorophyta	Green algae	7,000	
Phaeophyta	Brown algae	1,500	
Rhodophyta	Red algae	4,000	
Chrysophyta	Chrysophyte algae	12,500	
Pyrrophyta	Dinoflagellates	1,100	
Euglenophyta	Euglenoids	800	26,900
Plantae			
Bryophyta	Mosses, liverworts, hornworts	16,600	
Psilophyta	Psilopsids	9	
Lycopodiophyta	Lycophytes	1,275	
Equisetophyta	Horsetails	15	
Filicophyta	Ferns	10,000	
Gymnosperma	Gymnosperms	529	
Dicotolydonae	Dicots	170,000	
Monocotolydonae	Monocots	50,000	248,428

Table 1 *Numbers of Described Species of Living Organisms*[a] *(Continued)*

Kingdom and Major Subdivision	Common Name	No. of Described Species	Totals
Protozoa			
	Protozoans: Sarcomastigophorans, ciliates, and smaller groups	30,800	30,800
Animalia			
Porifera	Sponges	5,000	
Cnidaria, Ctenophora	Jellyfish, corals, comb jellies	9,000	
Platyhelminthes	Flatworms	12,200	
Nematoda	Nematodes (roundworms)	12,000	
Annelida	Annelids (earthworms and relatives)	12,000	
Mollusca	Mollusks	50,000	
Echinodermata	Echinoderms (starfish and relatives)	6,100	
Arthropoda	Arthropods		
Insecta	Insects	751,000	
Other arthropods		123,161	
Minor invertebrate phyla		9,300	989,761
Chordata			
Tunicata	Tunicates	1,250	
Cephalochordata	Acorn worms	23	
Vertebrata	Vertebrates		
Agnatha	Lampreys and other jawless fishes	63	
Chrondrichthyes	Sharks and other cartilaginous fishes	843	
Osteichthyes	Bony fishes	18,150	
Amphibia	Amphibians	4,184	
Reptilia	Reptiles	6,300	
Aves	Birds	9,040	
Mammalia	Mammals	4,000	43,853
TOTAL, all organisms			1,392,485

[a] *Compiled from multiple sources.*

to the same species or at least that they are capable of being linked genetically to them through chains of other breeding individuals. By definition they do not breed freely with members of other species.

This biological concept of species is the best ever devised, but it remains less than ideal. It works very well for most animals and some kinds of plants, but for some plant and a few animal populations in which intermediate amounts of hybridization occur, or ordinary sexual reproduction has been replaced by self-fertilization or parthenogenesis, it must be replaced with arbitrary divisions.

New species are usually created in one or the other of two ways. A large minority of plant species came into existence in essentially one step, through the process of polyploidy. This is a simple multiplication in the number of gene-bearing chromosomes—sometimes within a preexisting species and sometimes in hybrids between two species. Polyploids are typically not able to form fertile hybrids with the parent species. A second major process is geographic speciation and takes much longer. It starts when a single population (or series of populations) is divided by some barrier extrinsic to the organisms, such as a river, a mountain range, or an arm of the sea. The isolated populations then diverge from each other in evolution because of the inevitable differences of the environments in which they find themselves. Since all populations evolve when given enough time, divergence between all extrinsically isolated populations must eventually occur. By this process alone the populations can acquire enough differences to reduce interbreeding between them should the extrinsic barrier between them be removed and the populations again come into contact. If sufficient differences have accumulated, the populations can coexist as newly formed species. If those differences have not yet occurred, the populations will resume the exchange of genes when the contact is renewed.

Species diversity has been maintained at an approximately even level or at most a slowly increasing rate, although punctuated by brief periods of accelerated extinction every few tens of millions of years. The more similar the species under consideration, the more consistent the balance. Thus within clusters of islands, the numbers of species of birds (or reptiles, or ants, or other equivalent groups) found on each island in turn increases approximately as the fourth root of the area of the island. In other words, the number of species can be predicted as a constant X (island area)$^{0.25}$, where the exponent can deviate according to circumstances, but in most cases it falls between 0.15 and 0.35. According to this theory of island biogeography, in a typical case (where the exponent is at or near 0.25) the rule of thumb is that a ten-fold increase in area results in a doubling of a number of species (MacArthur and Wilson, 1967).

In a recent study of the ants of Hispaniola, I found fossils of thirty-seven genera (clusters of species related to each other but distinct from other such clusters) in amber from the Miocene age—about 20 million years old. Exactly thirty-seven genera exist on the island today. However, fifteen of the original thirty-seven have become extinct, while fifteen others not present in the Miocene deposits have invaded to replace them, thus sustaining the original diversity (Wilson, 1985b).

On a grander scale, families—clusters of genera—have also maintained a balance within the faunas of entire continents. For example, a reciprocal and apparently symmetrical exchange of land mammals between North and South America began 3 million years ago, after the rise of the Panamanian land bridge. The number of families in South America first rose from thirty-two to thirty-nine and then subsided to the thirty-five that exist there today. A comparable adjustment occurred in North America. At the generic level, North American elements dominated those from South America: twenty-four genera invaded to the south whereas only twelve invaded to the north. Hence, although equilibrium was roughly preserved, it resulted in a major shift in the composition of the previously isolated South American fauna (Marshall et al., 1982).

Each species is the repository of an immense amount of genetic information. The number of genes range from about 1,000 in bacteria and 10,000 in some fungi to 400,000 or more in many flowering plants and a few animals (Hinegardner, 1976). A typical mammal such as the house mouse (*Mus musculus*) has about 100,000 genes. This full complement is found in each of its myriad cells, organized from four strings of DNA, each of which comprises about a billion nucleotide pairs (George D. Snell, Jackson Laboratory, Maine, personal communication, 1987). (Human beings have genetic information closer in quantity to the mouse than to the more abundantly endowed salamanders and flowering plants; the difference, of course, lies in what is encoded.) If stretched out fully, the DNA would be roughly one meter long. But this molecule is invisible to the naked eye because it is only twenty angstroms in diameter. If we magnified it until its width equaled that of wrapping string, the fully extended molecule would be 960 kilometers long. As we traveled along its length, we would encounter some twenty nucleotide pairs or "letters" of genetic code per inch, or about fifty per centimeter. The full information contained therein, if translated into ordinary-size letters of printed text, would just about fill all fifteen editions of the *Encyclopaedia Britannica* published since 1768 (Wilson, 1985a).

The number of species and the amount of genetic information in a representative organism constitute only part of the biological diversity on Earth. Each species is made up of many organisms. For example, the 10,000 or so ant species have been estimated to comprise 10^{15} living individuals at each moment of time (Wilson, 1971). Except for cases of parthenogenesis and identical twinning, virtually no two members of the same species are genetically identical, due to the high levels of genetic polymorphism across many of the gene loci (Selander, 1976). At still another level, wide-ranging species consist of multiple breeding populations that display complex patterns of geographic variation in genetic polymorphism. Thus, even if an endangered species is saved from extinction, it will probably have lost much of its internal diversity. When the populations are allowed to expand again, they will be more nearly genetically uniform than the ancestral populations. The bison herds of today are biologically not quite the same—not so interesting—as the bison herds of the early nineteenth century.

THE NATURAL LONGEVITY OF SPECIES

Within particular higher groups of organisms, such as ammonites or fishes, species have a remarkably consistent longevity. As a result, the probability that a given species will become extinct in a given interval of time after it splits off from other species can be approximated as a constant, so that the frequency of species surviving through time falls off as an exponential decay function; in other words, the percentage (but not the absolute number) of species going extinct in each period of time stays the same (Van Valen, 1973).* These regularities, such as they are, have been interrupted during the past 250 million years by major episodes of extinction that have been recently estimated to occur regularly at intervals of 26 million years (Raup and Sepkoski, 1984).

Because of the relative richness of fossils in shallow marine deposits, the longevity of fish and invertebrate species living there can often be determined with a modest degree of confidence. During Paleozoic and Mesozoic times, the average persistence of most fell between 1 and 10 million years: that is, 6 million for echinoderms, 1.9 million for graptolites, 1.2 to 2 million for ammonites, and so on (Raup, 1981, 1984).

These estimates are extremely interesting and useful but, as paleontologists have generally been careful to point out, they also suffer from some important limitations. First, terrestrial organisms are far less well known, few estimates have been attempted, and thus different survivorship patterns might have occurred (although Cenozoic flowering plants, at least, appear to fall within the 1- to 10-million-year range). More important, a great many organisms on islands and other restricted habitats, such as lakes, streams, and mountain crests, are so rare or local that they could appear and vanish within a short time without leaving any fossils. An equally great difficulty is the existence of sibling species—populations that are reproductively isolated but so similar to closely related species as to be difficult or impossible to distinguish through conventional anatomical traits. Such entities could rarely be diagnosed in fossil form. Together, all these considerations suggest that estimates of the longevity of natural species should be extended only with great caution to groups for which there is a poor fossil record.

* Van Valen's original formulation, whose difficulties and implications are revealed by more recent research, has been discussed by Raup (1975) and by Lewin (1985). These studies deal with the clade, or set of populations descending through time after having split off as a distinct species from other such populations. They do not refer to the chronospecies, which is just a set of generations of the same species that is subjectively different from sets of generations.

RAIN FORESTS AS CENTERS OF DIVERSITY

In recent years, evolutionary biologists and conservationists have focused increasing attention on tropical rain forests, for two principal reasons. First, although these habitats cover only 7 percent of the Earth's land surface, they contain more than half the species in the entire world biota. Second, the forests are being destroyed so rapidly that they will mostly disappear within the next century, taking with them hundreds of thousands of species into extinction. Other species-rich biomes are in danger, most notably the tropical coral reefs, geologically ancient lakes, and coastal wetlands. Each deserves special attention on its own, but for the moment the rain forests serve as the ideal paradigm of the larger global crisis.

Tropical rain forests, or more precisely closed tropical forests, are defined as habitats with a relatively tight canopy of mostly broad-leaved evergreen trees sustained by 100 centimeters or more of annual rainfall. Typically two or more other layers of trees and shrubs occur beneath the upper canopy. Because relatively little sunlight reaches the forest floor, the undergrowth is sparse and human beings can walk through it with relative ease.

The species diversity of rain forests borders on the legendary. Every tropical biologist has a favorite example to offer. From a single leguminous tree in the Tambopata Reserve of Peru, I recently recovered forty-three species of ants belonging to twenty-six genera, about equal to the entire ant fauna of the British Isles (Wilson, 1987). Peter Ashton found 700 species of trees in ten selected one-hectare plots in Borneo, the same as in all of North America (Ashton, Arnold Arboretum, personal communication, 1987). It is not unusual for a square kilometer of forest in Central or South America to contain several hundred species of birds and many thousands of species of butterflies, beetles, and other insects.

Despite their extraordinary richness, tropical rain forests are among the most fragile of all habitats. They grow on so-called wet deserts—an unpromising soil base washed by heavy rains. Two-thirds of the area of the forest surface consists of tropical red and yellow earths, which are typically acidic and poor in nutrients. High concentrations of iron and aluminum form insoluble compounds with phosphorus, thereby decreasing the availability of phosphorus to plants. Calcium and potassium are leached from the soil soon after their compounds are dissolved from the rain. As little as 0.1 percent of the nutrients filter deeper than five centimeters beneath the soil surface (NRC, 1982). An excellent popular account of rain forest ecology is given by Forsyth and Miyata (1984).

During the 150 million years since its origin, the principally dicotyledonous flora has nevertheless evolved to grow thick and tall. At any given time, most of the nonatmospheric carbon and vital nutrients are locked up in the tissue of the vegetation. As a consequence, the litter and humus on the ground are thin compared to the thick mats of northern temperate forests. Here and there, patches of bare earth show through. At every turn one can see evidence of rapid decomposition by dense populations of termites and fungi. When the forest is cut and burned, the ash and decomposing vegetation release a flush of nutrients adequate to support new herbaceous and shrubby growth for two or three years. Then these materials decline to levels lower than those needed to support a healthy growth of agricultural crops without artificial supplements.

The regeneration of rain forests is also limited by the fragility of the seeds of the constituent woody species. The seeds of most species begin to germinate within a few days or weeks, severely limiting their ability to disperse across the stripped land into sites favorable for growth. As a result, most sprout and die in the hot, sterile soil of the clearings (Gomez-Pompa et al., 1972). The monitoring of logged sites indicates that regeneration of a mature forest might take centuries. The forest at Angkor (to cite an anecdotal example) dates back to the abandonment of the Khmer capital in 1431, yet is still structurally different from a climax forest today, 556 years later. The process of rain forest regeneration is in fact so generally slow that few extrapolations have been possible; in some zones of greatest combined damage and sterility, restoration might never occur naturally (Caufield, 1985; Gomez-Pompa et al., 1972).

Approximately 40 percent of the land that can support tropical closed forest now lacks it, primarily because of human action. By the late 1970s, according to estimates from the Food and Agricultural

Organization and United Nations Environmental Programme, 7.6 million hectares or nearly 1 percent of the total cover is being permanently cleared or converted into the shifting-cultivation cycle. The absolute amount is 76,000 square kilometers (27,000 square miles) a year, greater than the area of West Virginia or the entire country of Costa Rica. In effect, most of this land is being permanently cleared, that is, reduced to a state in which natural reforestation will be very difficult if not impossible to achieve (Mellilo et al., 1985). This estimated loss of forest cover is close to that advanced by the tropical biologist Norman Myers in the mid-1970s, an assessment that was often challenged by scientists and conservationists as exaggerated and alarmist. The vindication of this early view should serve as a reminder always to take such doomsday scenarios seriously, even when they are based on incomplete information.

A straight-line extrapolation from the first of these figures, with identically absolute annual increments of forest-cover removal, leads to A.D. 2135 as the year in which all the remaining rain forest will be either clear-cut or seriously disturbed, mostly the former. By coincidence, this is close to the date (2150) that the World Bank has estimated the human population will plateau at 11 billion people (World Bank, 1984). In fact, the continuing rise in human population indicates that a straight-line estimate is much too conservative. Population pressures in the Third World will certainly continue to accelerate deforestation during the coming decades unless heroic measures are taken in conservation and resource management.

There is another reason to believe that the figures for forest cover removal present too sanguine a picture of the threat to biological diversity. In many local areas with high levels of endemicity, deforestation has proceeded very much faster than the overall average. Madagascar, possessor of one of the most distinctive floras and faunas in the world, has already lost 93 percent of its forest cover. The Atlantic coastal forest of Brazil, which so enchanted the young Darwin upon his arrival in 1832 ("wonder, astonishment & sublime devotion, fill & elevate the mind"), is 99 percent gone. In still poorer condition—in fact, essentially lost—are the forests of many of the smaller islands of Polynesia and the Caribbean.

HOW MUCH DIVERSITY IS BEING LOST?

No precise estimate can be made of the numbers of species being extinguished in the rain forests or in other major habitats, for the simple reason that we do not know the numbers of species originally present. However, there can be no doubt that extinction is proceeding far faster than it did prior to 1800. The basis for this statement is not the direct observation of extinction. To witness the death of the last member of a parrot or orchid species is a near impossibility. With the exception of the showiest birds, mammals, or flowering plants, biologists are reluctant to say with finality when a species has finally come to an end. There is always the chance (and hope) that a few more individuals will turn up in some remote forest remnant or other. But the vast majority of species are not monitored at all. Like the dead of Gray's "Elegy Written in a Country Churchyard," they pass from the Earth without notice.

Instead, extinction rates are usually estimated indirectly from principles of biogeography. As I mentioned above, the number of species of a particular group of organisms in island systems increases approximately as the fourth root of the land area. This has been found to hold true not just on real islands but also on habitat islands, such as lakes in a "sea" of land, alpine meadows or mountaintops surrounded by evergreen forests, and even in clumps of trees in the midst of a grassland (MacArthur and Wilson, 1967).

Using the area-species relationship, Simberloff (1984) has projected ultimate losses due to the destruction of rain forests in the New World tropical mainland. If present levels of forest removal continue, the stage will be set within a century for the inevitable loss of 12 percent of the 704 bird species in the Amazon basin and 15 percent of the 92,000 plant species in South and Central America.

As severe as these regional losses may be, they are far from the worst, because the Amazon and Orinoco basins contain the largest continuous rain forest tracts in the world. Less extensive habitats are far more threatened. An extreme example is the western forest of Ecuador. This habitat was largely undisturbed until after 1960, when a newly constructed road network led to the swift incursion of settlers and clear-cutting of most of the area. Now only patches remain, such as the 0.8-square-kilometer tract at the Rio Palenque Biological Station. This tiny reserve contains 1,033 plant species, perhaps one-quarter of which are known only to occur in coastal Ecuador. Many are known at the present time only from a single living individual (Gentry, 1982).

In general, the tropical world is clearly headed toward an extreme reduction and fragmentation of tropical forests, which will be accompanied by a massive extinction of species. At the present time, less than 5 percent of the forests are protected within parks and reserves, and even these are vulnerable to political and economic pressures. For example, 4 percent of the forests are protected in Africa, 2 percent in Latin America, and 6 percent in Asia (Brown, 1985). Thus in a simple system as envisioned by the basic models of island biogeography, the number of species of all kinds of organisms can be expected to be reduced by at least one-half—in other words, by hundreds of thousands or even (if the insects are as diverse as the canopy studies suggest) by millions of species. In fact, the island-biogeographic projections appear to be conservative for two reasons. First, tropical species are far more localized than those in the temperate zones. Consequently, a reduction of 90 percent of a tropical forest does not just reduce all the species living therein to 10 percent of their original population sizes, rendering them more vulnerable to future extinction. That happens in a few cases, but in many others, entire species are eliminated because they happened to be restricted to the portion of the forest that was cut over. Second, even when a portion of the species survives, it will probably have suffered significant reduction in genetic variation among its members due to the loss of genes that existed only in the outer portions.

The current reduction of diversity seems destined to approach that of the great natural catastrophes at the end of the Paleozoic and Mesozoic eras—in other words, the most extreme in the past 65 million years. In at least one important respect, the modern episode exceeds anything in the geological past. In the earlier mass extinctions, which some scientists believe were caused by large meteorite strikes, most of the plants survived even though animal diversity was severely reduced. Now, for the first time, plant diversity is declining sharply (Knoll, 1984).

HOW FAST IS DIVERSITY DECLINING?

The area-species curves of island systems, that is, the quantitative relationship between the area of islands and the number of species that can persist on the islands, provide minimal estimates of the reduction of species diversity that will eventually occur in the rain forests. But how long is "eventually"? This is a difficult question that biogeographers have attacked with considerable ingenuity. When a forest is reduced from, say, 100 square kilometers to 10 square kilometers by clearing, some immediate extinction is likely. However, the new equilibrium will not be reached all at once. Some species will hang on for a while in dangerously reduced populations. Elementary mathematical models of the process predict that the number of species in the 10-square-kilometer plot will decline at a steadily decelerating rate, i.e., they will decay exponentially to the lower level.

Studies by Jared Diamond and John Terborgh have led to the estimation of the decay constants for the bird faunas on naturally occurring islands (Diamond, 1972, 1984; Terborgh, 1974). These investigators took advantage of the fact that rising sea levels 10,000 years ago cut off small land masses that had previously been connected to South America, New Guinea, and the main islands of Indonesia. For example, Tobago, Margarita, Coiba, and Trinidad were originally part of the South American mainland and shared the rich bird fauna of that continent. Thus they are called land-bridge islands. In a similar manner, Yapen, Aru, and Misol were connected to New Guinea. In the study of the South American land-bridge islands, Terborgh found that the smaller the island, the higher the estimated decay constant and hence extinction rate. Terborgh then turned to Barro Colorado Island, which was isolated for the first time by the rise of Gatun Lake during the construction of the Panama Canal. Applying the natural land-bridge extinction curve to an island of this size (seventeen square kilometers) and fitting the derived decay constant to the actual period of isolation (fifty years), Terborgh predicted an extinction of seventeen bird species. The actual number known to have vanished as a probable result of insularization is thirteen, or twelve percent of the 108 breeding species originally present. The extinction rates of bird species on Barro Colorado Island were based on careful studies by E. O. Willis and J. R. Karr and have been recently reviewed by Diamond (1984).

Several other studies of recently created islands of both tropical and temperate-zone woodland have produced similar results, which can be crudely summarized as follows: when the islands range from one to twenty-five square kilometers—the size of many smaller parks and reserves—the rate of extinction of bird species during the first 100 years is 10 to 50 percent. Also as predicted, the extinction rate is highest in the smaller patches, and it rises steeply when the area drops below one square kilometer. To take one example provided by Willis (1979), three patches of subtropical forest isolated (by agricultural clearing) in Brazil for about a hundred years varied from 0.2 to 14 square kilometers, and, in reverse order, their resident bird species suffered 14 to 62 percent extinction rates.

What do these first measurements tell us about the rate at which diversity is being reduced? No precise estimate can be made for three reasons. First, the number of species of organisms is not known, even to the nearest order of magnitude. Second, because even in a simple island-biogeographic system, diversity reduction depends on the size of the island fragments and their distance from each other—factors that vary enormously from one country to the next. Third, the ranges of even the known species have not been worked out in most cases, so that we cannot say which ones will be eliminated when the tropical forests are partially cleared.

However, scenarios of reduction can be constructed to give at least first approximations if certain courses of action are followed. Let us suppose, for example, that half the species in tropical forests are very localized in distribution, so that the rate at which species are being eliminated immediately is approximately this fraction multiplied by the rate-percentage of the forests being destroyed. Let us conservatively estimate that 5 million species of organisms are confined to the tropical rain forests, a figure well justified by the recent upward adjustment of insect diversity alone. The annual rate of reduction would then be $0.5 \times 5 \times 10^6 \times 0.007$ species, or 17,500 species per year. Given 10 million species in the fauna and flora of all the habitats of the world, the loss is roughly one out of every thousand species per year. How does this compare with extinction rates prior to human intervention? The estimates of extinction rates in Paleozoic and Mesozoic marine faunas cited earlier (Raup, 1981, 1984; Raup and Sepkoski, 1984; Van Valen, 1973) ranged according to taxonomic group (e.g., echinoderms versus cephalopods) from one out of every million to one out of every 10 million per year. Let us assume that on the order of 10 million species existed then, in view of the evidence that diversity has not fluctuated through most of the Phanerozoic time by a factor of more than three (Raup and Sepkoski, 1984). It follows that both the per-species rate and absolute loss in number of species due to the current destruction of rain forests (setting aside for the moment extinction due to the disturbance of other habitats) would be about 1,000 to 10,000 times that before human intervention.

I have constructed other simple models incorporating the quick loss of local species and the slower loss of widespread species due to the insularization effect, and these all lead to comparable or higher extinction rates. It seems difficult if not impossible to combine what is known empirically of the extinction process with the ongoing deforestation process without arriving at extremely high rates of species loss in the near future. Curiously, however, the study of extinction remains one of the most neglected in ecology. There is a pressing need for a more sophisticated body of theories and carefully planned field studies based on it than now exist.

WHAT CAN BE DONE?

The biological diversity most threatened is also the least explored, and there is no prospect at the moment that the scientific task will be completed before a large fraction of the species vanish. Probably no more than 1,500 professional systematists in the world are competent to deal with the millions of species found in the humid tropic forests. Their number may be dropping, due to decreased professional opportunities, reduced funding for research, and the assignment of a higher priority to other disciplines. Data concerning the number of taxonomists, as well as detailed arguments for the need to improve research in tropical countries, are given by NRC (1980). The decline has been accompanied by a more than 50 percent decrease in the number of publications in tropical ecology from 1979 to 1983 (Cole, 1984).

The problem of tropical conservation is thus exacerbated by the lack of knowledge and the paucity of ongoing research. In order to make precise assessments and recommendations, it is necessary to know which species are present (recall that the great majority have not even received a scientific name) as well as their geographical ranges, biological properties, and possible vulnerability to environmental change.

It would be a great advantage, in my opinion, to seek such knowledge for the entire biota of the world. Each species is unique and intrinsically valuable. We cannot expect to answer the important questions of ecology and other branches of evolutionary biology, much less preserve diversity with any efficiency, by studying only a subset of the extant species.

I will go further: the magnitude and control of biological diversity are not just central problems of evolutionary biology, but among the key problems of science as a whole. At present, there is no way of knowing whether there are 5, 10, or 30 million species on Earth. There is no theory that can predict what this number might turn out to be. With reference to conservation and practical applications, it also matters why a certain subset of species exists in each region of the Earth, and what is happening to each one year by year. Unless an effort is made to understand all of diversity, we will fall far short of understanding life in these important respects, and due to the accelerating extinction of species, much of our opportunity will slip away forever.

Lest this exploration be viewed as an expensive Manhattan Project unattainable in today's political climate, let me cite estimates I recently made of the maximum investment required for a full taxonomic accounting of all species: 25,000 professional lifetimes (4,000 systematists are at work full or part time in North America today); their final catalog would fill sixty meters of library shelving for each million species (Wilson, 1985a). Computer-aided techniques could be expected to cut the effort and cost substantially. In fact, systematics has one of the lowest cost-to-benefit ratios of all scientific disciplines.

It is equally true that knowledge of biological diversity will mean little to the vast bulk of humanity unless the motivation exists to use it. Fortunately, both scientists and environmental policy makers have established a solid linkage between economic development and conservation. The problems of human beings in the tropics are primarily biological in origin: overpopulation, habitat destruction, soil deterioration, malnutrition, disease, and even, for hundreds of millions, the uncertainty of food and shelter from one day to the next. These problems can be solved in part by making biological diversity a source of economic wealth. Wild species are in fact both one of the Earth's most important resources and the least utilized. We have come to depend completely on less than 1 percent of living species for our existence, the remainder waiting untested and fallow. In the course of history, according to estimates made by Myers (1984), people have utilized about 7,000 kinds of plants for food; predominant among these are wheat, rye, maize, and about a dozen other highly domesticated species. Yet there are at least 75,000 edible plants in existence, and many of these are superior to the crop plants in widest use. Others are potential sources of new pharmaceuticals, fibers, and petroleum substitutes. In addition, among the insects are large numbers of species that are potentially superior as crop pollinators, control agents for weeds, and parasites and predators of insect pests. Bacteria, yeasts, and other microorganisms are likely to continue yielding new medicines, food, and procedures of soil restoration. Biologists have begun to fill volumes with concrete proposals for the further exploration and better use of diversity, with increasing emphasis on the still unexplored portions of the tropical biota. Some of the most recent and useful works on this subject include those by Myers (1984), NRC (1975), Office of Technology Assessment (1984), Oldfield (1984), and the U.S. Department of State (1982). In addition, an excellent series of specialized publications on practical uses of wild species have been produced during the past ten years by authors and panels commissioned by the Board on Science and Technology for International Development (BOSTID) of the National Research Council.

In response to the crisis of tropical deforestation and its special threat to biological diversity, proposals are regularly being advanced at the levels of policy and research. For example, Nicholas Guppy (1984), noting the resemblance of the lumbering of rain forests to petroleum extraction as the mining of a nonrenewable resource for short-term profit, has recommended the creation of a cartel, the Organization of Timber-Exporting Countries (OTEC). By controlling production and prices of lumber, the organization could slow production while encouraging member states to "protect the forest environment in general and gene stocks and special habitats in particular, create plantations to supply industrial and fuel wood, benefit indigenous tribal forest peoples, settle encroachers, and much else." In another approach, Thomas Lovejoy (1984) has recommended that debtor nations with forest resources and other valuable habitats be given discounts or credits for undertaking conservation programs. Even a small amount of forgiveness would elevate the sustained value of the natural habitats while providing hard currency for alternatives to their exploitation.

Another opportunity for innovation lies in altering somewhat the mode of direct economic assistance to developing countries. A large part of the damage to tropical forests, especially in the New World, has resulted from the poor planning of road systems and dams. For example, the recent settlement of the state of Rondonia and construction of the Tucurui Dam, both in Brazil, are now widely perceived by ecologists and economists alike as ill-conceived (Caufield, 1985). Much of the responsibility of minimizing environmental damage falls upon the international agencies that have the power to approve or disapprove particular projects.

The U.S. Congress addressed this problem with amendments to the Foreign Assistance Act in 1980, 1983, and 1986, which call for the development of a strategy for conserving biological diversity. They also mandate that programs funded through the U.S. Agency for International Development (USAID) include an assessment of environmental impact. In implementing this new policy, USAID has recognized that "the destruction of humid tropical forests is one of the most important environmental issues for the remainder of this century and, perhaps, well into the next," in part because they are "essential to the survival of vast numbers of species of plants and animals" (U.S. Department of State, 1985). In another sphere, The World Bank and other multinational lending agencies have come under increasing pressure to take a more active role in assessing the environmental impact of the large-scale projects they underwrite (Anonymous, 1984).

In addition to recommendations for international policy initiatives, there has recently been a spate of publications on the linkage of conservation and economic use of tropical forests. Notable among them are *Research Priorities in Tropical Biology* (NRC, 1980), based on a study of the National Research Council; *Technologies to Sustain Tropical Forest Resources* (OTA, 1984), prepared by the Office of Technology Assessment for the U.S. Congress; and the *U.S. Strategy on the Conservation of Biological Diversity* (USAID, 1985), a report to Congress by an interagency task force. Most comprehensive of all—and in my opinion the most encouraging in its implications—is the three-part series *Tropical Forests: A Call for Action*, released by the World Resources Institute, The World Bank, and the United Nations Development Programme (1985). The report makes an assessment of the problem worldwide and reviews case histories in which conservation or restoration have contributed to economic development. It examines the needs of every tropical country with important forest reserves. The estimated cost to make an impact on tropical deforestation over the next five years would be U.S. $8 billion—a large sum but surely the most cost-effective investment available to the world at the present time.

In the end, I suspect it will all come down to a decision of ethics—how we value the natural worlds in which we evolved and now, increasingly, how we regard our status as individuals. We are fundamentally mammals and free spirits who reached this high a level of rationality by the perpetual creation of new options. Natural philosophy and science have brought into clear relief what might be the essential paradox of human existence. The drive toward perpetual expansion—or personal freedom—is basic to the human spirit. But to sustain it we need the most delicate, knowing stewardship of the living world that can be devised. Expansion and stewardship may appear at first to be conflicting goals, but the opposite is true. The depth of the conservation ethic will be measured by the extent to which each of the two approaches to nature is used to reshape and reinforce the other. The paradox can be resolved by changing its premises into forms more suited to ultimate survival, including protection of the human spirit. I recently wrote in synecdochic form about one place in South America to give these feelings more exact expression:

> To the south stretches Surinam eternal, Surinam serene, a living treasure awaiting assay. I hope that it will be kept intact, that at least enough of its million-year history will be saved for the reading. By today's ethic its value may seem limited, well beneath the pressing concerns of daily life. But I suggest that as biological knowledge grows the ethic will shift fundamentally so that everywhere, for reasons that have to do with the very fiber of the brain, the fauna and flora of a country will be thought part of the national heritage as important as its art, its language, and that astonishing blend of achievement and farce that has always defined our species (Wilson, 1984).

References

Anonymous. 1984. Critics fault World Bank for ecological neglect. Conserv. Found. News. Nov.-Dec.:1–7.

Arnett, R. H. 1985. General considerations. Pp. 3–9 in American Insects: A Handbook of the Insects of America North of Mexico. Van Nostrand Reinhold, New York.

Brown, R. L., ed. 1985. State of the World 1985: A Worldwatch Institute Report on Progress Toward a Sustainable Society. W. W. Norton, New York. 301 pp.

Caufield, C. 1985. In the Rainforest. A. A. Knopf, New York. 283 pp.

Cole, N. H. A. 1984. Tropical ecology research. Nature 309:204.

Diamond, J. M. 1972. Biogeographic kinetics: Estimation of relaxation times for avifaunas of Southwest Pacific islands. Proc. Natl. Acad. Sci. USA 69:3199–3203.

Diamond, J. M. 1984. "Normal" extinctions of isolated populations. Pp. 191–246 in M. H. Nitecki, ed. Extinctions. University of Chicago Press, Chicago.

Erwin, T. L. 1983. Beetles and other insects of tropical forest canopies at Manaus, Brazil, sampled by insecticidal fogging. Pp. 59–75 in S. L.

Sutton, T. C. Whitmore, and A. C. Chadwick, eds. Tropical Rain Forest: Ecology and Management. Blackwell, Edinburgh.

Forsyth, A., and K. Miyata. 1984. Tropical Nature: Life and Death in the Rain Forests of Central & South America. Scribner's, New York. 272 pp.

Frankel, O. H., and M. E. Soulé. 1981. Conservation and Evolution. Cambridge University Press, Cambridge, Mass. 327 pp.

Gentry, A. H. 1982. Patterns of Neotropical plant-species diversity. Evol. Biol. 15:1–85.

Gomez-Pompa, A., C. Vazquez-Yanes, and S. Guevara. 1972. The tropical rain forest: A nonrenewable resource. Science 177:762–765.

Guppy, N. 1984. Tropical deforestation: A global view. Foreign Affairs 62:928–965.

Hinegardner, R. 1976. Evolution of genome size. Pp. 179–199 in F. J. Ayala, ed. Molecular Evolution. Sinauer Associates, Sunderland, Mass.

Knoll, A. H. 1984. Patterns of extinction in the fossil record of vascular plants. Pp. 21–68 in M. H. Nitecki, ed. Extinctions. University of Chicago Press, Chicago.

Lewin, R. 1985. Red Queen runs into trouble? Science 227:399–400.

Lovejoy, T. E. 1984. Aid debtor nations' ecology. New York Times, October 4.

MacArthur, R. H., and E. O. Wilson. 1967. The Theory of Island Biogeography. Princeton University Press, Princeton, N.J. 203 pp.

Marshall, L. G., S. D. Webb, J. J. Sepkoski, Jr., and D. M. Raup. 1982. Mammalian evolution and the great American interchange. Science 215:1351–1357.

Melillo, J. M., C. A. Palm, R. A. Houghton, G. M. Woodwell, and N. Myers. 1985. A comparison of two recent estimates of disturbance in tropical forests. Environ. Conserv. 12:37–40.

Myers, N. 1983. A Wealth of Wild Species: Storehouse for Human Welfare. Westview Press, Boulder, Colo. 300 pp.

Myers, N. 1984. The Primary Source: Tropical Forests and Our Future. W. W. Norton, New York. 399 pp.

NRC (National Research Council). 1975. Underexploited Tropical Plants with Promising Economic Value. Board on Science and Technology for International Development Report 16. National Academy of Sciences, Washington, D.C. 187 pp.

NRC (National Research Council). 1979. Tropical Legumes: Resources for the Future. Board on Science and Technology for International Development Report 25. National Academy of Sciences, Washington, D.C. 331 pp.

NRC (National Research Council). 1980. Research Priorities in Tropical Biology. National Academy of Sciences, Washington, D.C. 116 pp.

NRC (National Research Council). 1982. Ecological Aspects of Development in the Humid Tropics. National Academy Press, Washington, D.C. 297 pp.

Oldfield, M. L. 1984. The Value of Conserving Genetic Resources. National Park Service, U.S. Department of the Interior, Washington, D.C. 360 pp.

OTA (Office of Technology Assessment). 1984. Technologies to Sustain Tropical Forest Resources. Congress of the United States, Office of Technology Assessment, Washington, D.C. 344 pp.

Parker, S. P., ed. 1982. Synopsis and Classification of Living Organisms. McGraw-Hill, New York. 2 vols.

Raup, D. M. 1975. Taxonomic survivorship curves and Van Valen's Law. Paleobiology 1:82–86.

Raup, D. M. 1981. Extinction: Bad genes or bad luck? Acta Geol. Hisp. 16:25–33.

Raup, D. M. 1984. Evolutionary radiations and extinction. Pp. 5–14 in H. D. Holland and A. F. Trandell, eds. Patterns of Change in Evolution. Dahlem Konferenzen, Abakon Verlagsgesellschaft, Berlin.

Raup, D. M., and J. J. Sepkoski, Jr. 1984. Periodicity of extinctions in the geologic past. Proc. Natl. Acad. Sci. USA 81:801–805.

Selander, R. K. 1976. Genic variation in natural populations. Pp. 21–45 in F. J. Ayala, ed. Molecular Evolution. Sinauer Associates, Sunderland, Mass.

Simberloff, D. S. 1984. Mass extinction and the destruction of moist tropical forests. Zh. Obshch. Biol. 45:767–778.

Terborgh, J. 1974. Preservation of natural diversity: The problem of extinction-prone species. BioScience 24:715–722.

USAID (U.S. Agency for International Development). 1985. U.S. Strategy on the Conservation of Biological Diversity. An Interagency Task Force Report to Congress. U.S. Agency for International Development, Washington, D.C. 52 pp.

U.S. Department of State. 1982. Proceedings of the U.S. Strategy Conference on Biological Diversity. November 16–18, 1981, Washington, D.C. Publication No. 9262. U.S. Department of State, Washington, D.C. 126 pp.

U.S. Department of State. 1985. Humid Tropical Forests: AID Policy and Guidance. U.S. Department of State Memorandum. Government Printing Office, Washington, D.C. 3 pp.

Van Valen, L. 1973. A new evolutionary law. Evol. Ther. 1:1–30.

Willis, E. O. 1979. The composition of avian communities in remanescent woodlots in southern Brazil. Papeis Avulsos Zool. 33:1–25.

Wilson, E. O. 1971. The Insect Societies. Belknap Press of Harvard University Press, Cambridge, Mass. 548 pp.

Wilson, E. O. 1984. Biophilia. Harvard University Press, Cambridge, Mass. 176 pp.

Wilson, E. O. 1985a. The biological diversity crisis: A challenge to science. Issues Sci. Technol. 2:20–29.

Wilson, E. O. 1985b. Invasion and extinction in the West Indian ant fauna: Evidence from the Dominican amber. Science 229:265–267.

Wilson, E. O. 1987. The arboreal ant fauna of Peruvian Amazon forests: A first assessment. Biotropica 2:245–251.

World Bank. 1984. World Development Report 1984. Oxford University Press, New York. 286 pp.

World Resources Institute, The World Bank, and United Nations Development Programme. 1985. Tropical Forests: A Call for Action. World Resources Institute, Washington, D.C. 3 vols.

Do We Really Want Diversity?

Reed F. Noss

"Managing for diversity" is the code of today's land managers, but in many cases "managing for weeds" would be a more accurate description of what actually goes on in the field. Our love for diversity, friends, is an ecological trap.

Conservationists often speak loftily of preserving "biological diversity" and "genetic diversity" as if the meaning and application of these concepts were self-evident. In reality, the scale and content of biological diversity are often far from clear.

More than one conservationist has been horribly surprised when the concept of diversity has been used against him by those who would convert our last natural areas into economic production units. The U.S.D.A. Forest Service is presently preparing land and resource management plans for all of the national forests, and one would hope that maintaining a diversity of wildlife in the forests would be a major objective of these plans. The Forest Service says it is. But curiously, the "preferred alternative," which invariably calls

Whole Earth Review, Summer 1987.

for more roads, more intensive silviculture, and increased timber harvest, is also considered to do the most for wildlife diversity.

How could this be? Is the Forest Service lying to us? In this case, probably not. When a forest is fragmented by roads and clearcuts, the resulting patchwork of habitats is almost always richer in species than the original, unfragmented forest. In addition to climax forest species (many but not all of which dwindle away after fragmentation), species dependent on early successional habitats often thrive under intensive forest management regimes. This is the perverse logic of the maximum diversity concept: bring in humans, roads, and machines, rip apart the old growth, and we will have all the more species. Good old human progress and wildlife working together!

But wait, friends, the story is much more complicated than the Forest Service and other manipulative land managers would have us believe. We cannot deny that human disturbance will often increase the *number* of species within single management units or even entire forests. But what about the *identity* of those species? The kinds of species that benefit from human disturbance are primarily plant and animal weeds. They are opportunistic generalists that get along just fine in the human-dominated agricultural and urban landscapes that surround our remaining natural areas. Opportunistic weeds do not need forests, parks, or preserves for survival.

On the other hand, the species that disappear from fragmented and human-disturbed forests are those we can least afford to lose. These are wilderness species, wide-ranging animals with large area requirements, and other organisms sensitive to the intrusions of men and machines. These sensitive species cannot usually survive without large nature reserves.

Examples of weedy species proliferating in disturbed areas and increasing overall diversity abound. A recent study in the New Jersey Pine Barrens focused on the effects of water pollution derived from residential and agricultural development. More species of aquatic macrophytes (vascular plants) were found in the polluted sites than in the unpolluted sites. But guess what? The polluted sites were dominated by marginal or nonindigenous species that are common to wetlands throughout the eastern United States. The unpolluted sites—though less diverse—contained a unique and distinctive Pine Barrens flora that is disappearing as land is developed in the region.

Human trampling in the vicinity of trails is another diversifying factor. Many studies have documented that trails create new microhabitats in their vicinity, leading to an increase in the number of plant species. Many of the new species probably "hitch-hiked" in as burrs on the pantlegs of hikers. But what about the rare and attractive orchids that are plucked by hikers who gained access by the trail? Are we willing to trade one rare orchid species for a dozen cosmopolitan weeds? And what about the animals that are disturbed by the frequent presence of hikers along the trail? In conservation, it is generally a mistake to treat all species as equal. We must focus on those species that suffer most from human disturbances.

The notorious edge effect is a classic example of the maximum-diversity concept gone awry. Wildlife biologists early in this century (particularly my ideological hero, Aldo Leopold) noticed that edges—the places where distinct habitats meet—are often richer in species than either of the adjoining habitats. This was explained by observations that edges contain animals from both of the adjoining habitats, animals that need both kinds of habitat for their life functions, and other animals that actually specialize on edges. Edges were found to be especially productive of certain favored games species like rabbits, pheasant, and bobwhite quail.

With all these tantalizing benefits of edges before their eyes, wildlife managers set out to create as much edge habitat as they possibly could. "Managing for diversity" usually meant managing for edge. But not everyone can win with such a management regime. Edge species tend to be weeds, and the species of habitat interiors tend to disappear when habitat area is reduced to favor high edge-interior ratios in management units.

Edge effects include even more insidious processes. Edge habitat, often drier and denser than interior habitat, typically extends a considerable distance into the forest interior. Weedy species then invade from the edges to alter species composition throughout a small forest block. Forest birds suffer reduced reproductive success when nest predators (for example, grackles, jays, crows, and small mammals) and brood parasites (brown-headed cowbirds) move in from the edges. People and their domestic animals also invade natural areas from their perimeters.

A study I conducted in an Ohio nature reserve surrounded by suburbs and agricultural land found an extraordinarily high diversity of breeding birds. Unfortunately, the dominant species in this 500-acre reserve were the same ones that dominated the surrounding developed land. Typical forest interior birds of the region had very small populations in the reserve and were in danger of local extinction. Management for habitat diversity within the reserve (especially the maintenance of early successional habitats and numerous edges) intensified the biological deterioration.

Disturbance, of course, is fundamentally a natural phenomenon that provides suitable niches for a variety of native flora and fauna. Fire, windthrow, floods, landslides, and other natural disturbance events are responsible for maintaining environmental heterogeneity at multiple scales. Many wildlife species are dependent upon the early successional habitats created by disturbance for food and other critical needs. Some native species are "fugitives" that cannot compete in climax communities and survive only by dispersing among recently disturbed patches. Even the climax forest is diversified by small-scale disturbances such as treefalls. Many of the tree species we associate with old-growth habitat actually require multiple treefall gap episodes in their vicinity in order to reach maturity.

But the ecological mosaic created by natural disturbance is a far cry from the checkerboard of isolated habitats created by modern humans. The natural mosaic is interconnected—the artificial patchwork is fragmented. This is an important distinction for species that require large systems of continuous habitat for survival. Additionally, artificial habitat manipulation generally requires roads for access. Nothing is worse for sensitive wildlife than a road. Roads bring vehicles, guns, noise, and weeds. A bear (Smokey notwithstanding) can usually deal with fire, windthrow, and flood—but he is in big trouble when surrounded by drunken poachers with guns, dogs, and citizens' band radios.

The critical point in all this is that the diversity concept does not prescribe straightforward recommendations for conservation. A more diverse system, in terms of number of species or habitats, is not necessarily more valuable than a simpler system. A relatively depauperate system may be the natural system for the area of concern. Another important consideration is scale. Manipulative management for edge and habitat interspersion may *increase* the number of species at the scale of an individual forest or nature reserve, but *decrease* the number of species in the biogeographical region. This switch occurs when the managed area simply perpetuates those species that are common in the developed landscape, while the species most in need of reserves for survival are lost from the region. Species dependent upon large blocks of unfragmented habitat—wilderness—are the first to disappear.

If we carry this fragmentation process to its logical extreme, we end up with a bland and boring biosphere composed only of opportunistic weeds, those species that can adapt readily to human development. Eventually, every place of similar climate ends up with virtually the same set of cosmopolitan species. Local character disappears. Diversification tragically becomes homogenization.

Ecologists are becoming aware of these diversity problems. But many foresters, wildlife biologists, park managers, and otherwise knowledgeable naturalists are being sucked into the trap of maximum diversity. Conservationists have been fooled and confused about what diversity means in context. They are unable to argue with the Forest Service's management plans that ostensibly maximize both hard commodities and wildlife. They are unaware of the divergent effects that a land management regime can have at different spatial and temporal scales. They think they are getting diversity, but they are really getting impoverishment.

I believe that conservationists do and should want diversity. We should strive to maintain every species in its rightful place on this Earth, and furthermore assure each species the potential to evolve as conditions change. But for any given area, the number of species or habitats alone is a poor criterion for conservation. What we want is the full complement of native species in natural or normal patterns of abundance. Call that native diversity. And tell the land managers about it.

The Future Is Today

Chris Maser

For Ecologically Sustainable Forestry

INTRODUCTION

As we human beings lost our spiritual connection with the Earth, as we lost the inner ground of our being, of our place in the world, we lost sight of the reciprocal interrelatedness of all life. We now walk the Earth with impoverished souls. Human poverty, global deforestation, global degradation and poisoning of the land, and the nuclear arms race are but a few metaphors of this emptiness of soul. Unless our minds and hearts are set on maintaining a biologically sustainable forest, each succeeding generation will have less than the preceding one, and their choices for survival will be equally diminished.

Trumpeter, vol. 7, no. 2, Spring 1990.

POVERTY AS METAPHOR

Zane, my wife, and I recently became the sponsors of two children in Ecuador, a small country in South America sandwiched between Colombia and Peru. Ecuador is about the size of the state of Nevada, but whereas Nevada is about 87 percent public lands and has relatively few permanent residents, Ecuador has a population of approximately 9.6 million people that is projected to double to 19.2 million people by 1993. Thus, poverty becomes the ever-present metaphor of the human condition.

Our new children, Patricia and Eduardo, were born into abject poverty. Six-year-old Patricia is undernourished and in poor health. She has six brothers and sisters, and although both her father, age 50, and her mother, age 36, work as day laborers, the family income is about 68 U.S. dollars per month. Their house is made of split cane, with a tin roof and a floor of rough boards. There are two rooms and two beds for the entire family. They have water, electricity, a portable gas stove for cooking, but no latrine.

Eduardo is 4 and is in good health. He has a brother and a sister. His mother, age 25, works as a washer woman; his father has abandoned the family. In addition to Eduardo, his brother and sister, and his mother, a grandmother, age 51, who takes care of the home and the youngsters, a grandfather, age 53, and an aunt, age 27, all live under the same roof. Although the grandfather is a street vendor and the aunt is a factory worker, the pooled family income is about 32 U.S. dollars per month. Their house too is of split cane with a tin roof and a floor of rough boards. There are two rooms and three beds. They have water, electricity, and a latrine, and they cook with firewood.

I remember working in North Africa and in Asia and seeing this kind of poverty, but that was over twenty years ago and time has dimmed my memory. In addition, I was younger and did not see quite so clearly as I do today. When the information about our new children arrived and I read it, all I could do was sit in stunned inner silence as the years in Africa and Asia came flooding back to engulf me. For the first time, I had an insight into real poverty. Oh, I had lived amongst it and had looked at it, but somehow I had not really seen it. Now, with our new children, it came home to me with searing clarity how much difference a mere 12 dollars per month can make. It was giving our children a choice they otherwise would not have . . . a choice that equals hope. In turn, choice and hope equal dignity . . . the birthright of every human being. How little for us to give from our incredible material wealth as a middle-class American family, and yet how much for them to receive, for to these children ours is not a gift of money but rather a gift of hope.

It occurs to me, thinking about our new children, that as human beings, all we have to bequeath future generations is options, choices, and each option, each choice, represents the future's limitations. Future generations must respond to our momentary decisions—including the present struggle over the ancient forests—which will become their inherited circumstances, and because the future *must* respond to our decisions as their circumstances, the future is today.

WHY SAVE THE ANCIENT FOREST?

There are many valid reasons to save ancient forests from extinction, as many, perhaps, as there are for saving tropical forests. One is that our forests of the Pacific Northwest are beautiful and are unique in the world (Waring and Franklin, 1979). Another is that the old trees of the Pacific Northwest inspire spiritual renewal in many people and are among the rapidly dwindling living monarchs of the world's forests. They are unique, irreplaceable, and finite in number, and they shall exist precisely once in forever. We can perhaps grow large trees over two or three centuries, but no one has ever done that on purpose. And if they did, such trees will not be Nature's trees; they will be humanity's trees. And although they may be just as beautiful as those created by Nature, they will be different in the human mind. A third reason is that a number of organisms, such as the spotted owl (*Strix occidentalis*) and the flying squirrel (*Glaucomys sabrinus*), either find their optimum habitat in these ancient forests or require the structures provided by the old trees, such as large snags and large fallen trees, to survive (Forsman et al., 1984; Franklin et al., 1981; Harris, 1984; Meslow et al., 1981). And a fourth reason is that ancient forests are the only living laboratories through which we and the future may be able to learn how to create sustainable forests—something no one in the world has so far accomplished.

As a living laboratory, ancient forests serve four vital functions. First, they are our link to the past, to the historical forest. The historical view tells us what the present is built on, and together the past and the present tell us what the future is projected on. Because the whole forest cannot be seen without taking long views both into the future and into the past, to lose the ancient forests is to cast ourselves adrift in a sea of almost total uncertainty with respect to the creation and sustainability of future forests and tree farms. We must remember that knowledge is only in past tense; learning is only in present tense; and prediction is only in future tense. To have sustainable forests and tree farms, we need to be able to know, to learn, and to predict. Without ancient forests, we eliminate learning, limit our knowledge, and greatly diminish our ability to predict.

Second, we did not design the forest, so we do not have a blueprint, parts catalog, or maintenance manual with which to understand and repair it. Nor do we have a service department in which the necessary repairs can be made. Therefore, how can we afford to liquidate the ancient forest that acts as a blueprint, parts catalog, maintenance manual, and service station—our only hope of understanding the sustainability of our simplified, plantation-mode tree farms?

Third, we are playing "genetic roulette" with tree farms of the future. What if our genetic engineering, our genetic cloning, our genetic streamlining, our genetic simplifications run amuck, as they so often have around the world? Native forests, be they ancient or young, are thus imperative because they—and only they—contain the entire genetic code for living, healthy, adaptable forests.

Fourth, intact segments of the ancient forest from which we can learn will allow us to make the necessary adjustments in both our thinking and our subsequent course of management to help assure the sustainability of forests and tree farms. If we choose not to deal with the heart of the ancient forest issue—sustainable forestry, we will find that reality is more subtle than our understanding of it and that our "good intentions" will likely give bad results.

Although there are many valid reasons to save ancient forests, there is only one reason that I know of for liquidating them—short-term economics. Economics, however, is the common language of Western civilization; is it not therefore wise to carefully consider whether saving substantial amounts of well-distributed ancient forests is a necessary part of the equation for maintaining a solvent forest industry?

Can we really afford to liquidate our remaining ancient forests? I have often heard that "we can't afford to save 'old-growth'; it's too valuable and too many jobs are at stake." I submit, however, that we must be exceedingly cautious lest economic judgment isolate us from the evidence that without biologically sustainable forests, we will not have an economically sustainable forest industry, and without an economically sustainable forest industry, there will be human communities in which we cannot have a sustainable economy. Therefore, if we liquidate the ancient forests of the Pacific Northwest—our living laboratories of which the spotted owl is the designated symbol—and our tree farms fail, as tree farms are failing over much of the world, we will be further impoverishing our souls and those of future generations through the myopic worship of the golden idol of materialism: short-term profits.

IF WE REALLY WANT THE SPOTTED OWL TO SURVIVE

The northern spotted owl has become the chosen symbol for the ancient forest, a symbol in the struggle of conflicting values—short-term economics versus all other human and other values of ancient forests. The spotted owl is the symbol for the survival of the ancient forest, but what does it really symbolize? The spotted owl is called an "indicator species" because its presence supposedly indicates a healthy ancient forest, but what does it really indicate? The spotted owl may be seen as a symbol for the survival of ancient forests, but in reality it is an indicator species for the planned extinction of ancient forests. Although the spotted owl was selected as the symbol of the ancient forest with good intentions, the results are bad if the objective is to save the owl because the real issue is not the owl. The real issue is the economics of extinction (Chasan, 1977)—the planned liquidation of ancient forests for short-term economic gains.

If we really want spotted owls to survive, then we must want ancient forests to survive also because on the scientific side there is evidence beyond a reasonable doubt that northern spotted owls require the unique structural components of ancient forests (Forsman et al., 1984; Franklin et al., 1981; Irwin, 1986). Where is the scientific data to the contrary? There is none that I know of. And yet, set-asides of ancient forest, as now planned, will create self-destruct islands of time-limited old-growth trees in a sea of young-growth tree farms. Unless a portion of the existing mature forest also is set aside to replace the ancient forest as it falls apart with age or by unplanned catastrophe, the spotted owl is doomed. And then the only difference is time if a portion of the young-growth forest is not also committed to replace the mature forest, as needed to maintain quality spotted owl habitat. We are planning spotted owl habitat in terms of absolute minimums and the liquidation of the ancient forest in terms of flexible maximums.

The great historian, Arnold Toynbee (1958), asked the critical question of why twenty-six great civilizations fell. He concluded that *they could not or would not change their direction, their way of thinking, to meet the changing conditions of life*. On page 298, Toynbee says of history, "We cannot say [what will happen] since we cannot foretell the future. We can only see that something which has actually happened once, in another episode of history, must at least be *one of the possibilities* [emphasis mine] that lie ahead of us." In addition, a knowledge of unwise historical choices and their disastrous consequences not only gives us the option of doing something different, but also gives us the possibility of altering an outcome in the future. Herein lies the hope of humanity. But to grasp that hope for the children of today and of tomorrow and beyond, we must ask ourselves, "Can society afford the environmental costs of the economics of extinction?" Have we become so myopic in our economic view that we are willing to risk losing the ability to have sustainable forests by pursuing the short-sighted, short-term, economic windfall to be had by cutting the remaining ancient forests?

In this context, it is imperative to understand that humanity has not "reforested" a single acre, because no one has planted and grown a forest on purpose. What we and the rest of the world have done, and are doing under the guise of "forestry," is trade our forests in on simplistic, economic tree farms. And forests and tree farms are not synonymous, no matter how many high-priced public relations firms try to create the impression that they are.

A FOREST VERSUS A TREE FARM

Before we can discuss a forest versus a tree farm, we must "define" forest and forestry and a tree farm and tree-farm management. A *forest* is the most complex, terrestrial biotic portion of the ecosystem, and is characterized by a predominance of trees. *Forestry* is the profession that embraces the science, art, and business of managing the forested portion of the ecosystem in a manner that assures the maintenance and sustainability of biological diversity and productivity for perpetual production of amenities, services, and goods for human use. A *tree farm* is an area under cultivation, a group of cultivated trees. And because a plantation is an economic crop, it is grossly simplified and specialized. *Tree-farm management* is the profession that embraces the science, art, and business of managing a tree farm—an agricultural crop—to reap the greatest economic returns on the least economic investment, in the shortest possible time.

Today's "forest practices" are counter to sustainable forestry because, instead of training foresters to manage forests, we train tree-farm managers to manage the short-rotation, "economic" tree farms with which we are replacing our native forests. Forests have evolved through the cumulative addition of structural diversity that initiates and maintains process diversity, complexity, and stability through time. We are reversing the rich building process of that diversity, complexity, and stability by replacing native forests with tree farms designed only with narrow, short-term economic considerations.

Thus, every acre on which a forest is replaced with a tree farm is an acre that is purposefully stripped of its biological diversity, of its biological sustainability, and is purposefully reduced to the lowest common denominator possible—simplistic economics. Simplistic economic theory in agriculture has not proven to be biologically sustainable anywhere in the world. So the concept of a "tree farm" is a strictly simplistic economic concept that has nothing to do with the biological sustainability of a forest. Under this concept, native forests are being replaced more and more with tree farms of genetically manipulated trees accompanied by the corporate-political-academic promise that such tree farms are better, healthier, and more viable than are the native forests that evolved with the land over millennia.

NATIVE FORESTS

In our burgeoning, product-oriented society, one of the most insidious dangers to native forest—that which has experienced no disruptive, human intrusion—be they ancient or young, is the perceived lack of value in maintaining an area for its *potential*. By that I mean its research value as an ecological blueprint of what a biologically sustainable forest is and how it functions, educational value, spiritual value, or any other value that does not turn an immediate, visible, economic profit. This short-sightedness is understandable considering that: (1) the Native North Americans viewed the land and all it contained as a "Thou," which is holy and is to be revered; Europeans, on the other hand, viewed the same land and all it contained—including the indigenous peoples—as an "it," which is simply an object to be exploited (Buber, 1970; Campbell with Moyers, 1988); (2) we in Western civilization focus predominantly on utilizable products from the ecosystem, rather than on the processes that produce the products; (3) renewable "natural" products are largely manifested above ground, whereas many of the processes that produce the products are below ground; (4) we therefore think about and manage what is visible above ground, and tend to ignore the processes below the surface of the soil; and (5) short-term economics is the driving force behind management of renewable natural resources and our society.

When these points are taken together, they form the foundation of Western economic culture. Reared with this historical background, it is difficult for most people to really understand the risks involved to humanity's future by violating all remaining natural areas either in principle or in fact. Although this may seem a bold statement, consider that, in addition to representing a collection of native species of both plants and animals with X amount of genetic diversity, each protected area of native forest, whether ancient or young, also represents a repository with a portion of the world's ecological processes and functions in living laboratories.

For example, the native temperate coniferous forests of the Pacific Northwest are still "healthy," whereas both the temperate coniferous forests and tree farms of central Europe are dying, and far more is known scientifically about forests of the Pacific Northwest and how they function than is known about European forests (Franklin et al., 1981; Harmon et al., 1986; Harris, 1984; Maser et al., 1988; and others). So a system of natural areas of native forest in the healthy Douglas-Fir (*Pseudotsuga menziesii*) region of the Pacific Northwest is a repository for ecological processes that, although different in specifics, are similar in principles to those of the dying forests and the dying Norway spruce (*Picea abies*) tree farms of central Europe (Blaschke and Baumler, 1986; Durrieu et al., 1984). By analogy, rather than a historical transplant of a particular species to reintroduce it into an area from which it has been extirpated, we have the potential to perform "global process-information transplants" through ecological knowledge that is and can be gleaned from and maintained through benchmark areas that represent Nature's blueprint—natural areas—whether or not today economists and industrialists can see anything but "economic waste" in saving them.

Areas of native ancient forest are even more important now than ever before, because today we face a generalized global warming called the "greenhouse effect." Such a historically unprecedented warming means that forests must be adaptable—both plants and animals, which constitute the interrelated, often symbiotic, biological processes of life. The problem for humanity is that no one knows which species will be able to adapt to such changes. What is known, however, is that native species are much more likely to be able to adapt than exotics, even of the same species, brought in from other areas. In this sense, adaptability equates to resilience in the face of sudden, dramatic, perhaps irreversible change.

Part of the process of maintaining ecological resilience is setting aside an ecologically adequate system of natural areas of native, ancient forest—an unconditional gift of potential knowledge for the future. In so doing, present and future generations have a repository not only of species, which more often than not are region-specific, but also of processes, which more often than not are world-wide in principle and application.

From such repositories, in addition to monitoring human-caused changes and maintaining habitat for particular species, it will be possible to learn how to maintain, restore, and sustain biological processes in various portions of the ecosystem. In this sense, reserves of native, ancient forests are the parts catalogs and maintenance manuals not only for that which is but also for that which can be.

Whatever we do to move toward biologically sustainability forestry—and we *must* act—will take the utmost in courage. With the right attitude, any mistakes we make may become the future's strength. But we must act while the Earth still has the strength and the resources to survive in the face of ongoing errors, and while there still is the ecological margin to allow a few more mistakes from which to learn. To assure the future's potential to correct our errors and its people's ability to learn from them, we must remember always that all we have to give the future is options and that an option spent—liquidating the ancient forest for whatever rationalized reason—is no longer an option. Therefore, we must ask, each time we make a decision that deals with natural resources, "How will our decision either maintain or enhance the options for the future?" That is our moral responsibility as human beings, because all we have to bequeath future generations is options, and each option foreclosed represents the limitation—the impoverishment—of the future. The generations to come have no choice but to respond to our decisions, which will have become their inherited circumstances, and because the decisions we are making today are inexorably creating the circumstances of the future, in sober reality the future is today.

CONCLUSION

As human beings, we participate in creation of the world we live in, because our very existence and that of every other living thing is involved in this on-going act of creation. As conscious, co-creators, we are the moral, ecological guides for the future. In this sense, our impressions of our ancestors are reflections of the care they took of the land that we inherited, and as the ancestors of future generations, we know that their impressions of us will be mirrored in the care we take of the land that they must inherit. Thus, if we would change our image, we must begin now, consciously, to create a new paradigm for our trusteeship of the land, one based on a sense of place and permanence, a sense of creation and landscape artistry, a sense of ecological health and sustainability, and a sense of humility and humanity. Although such a harmonious union between people and the Earth is not new in the world, it is new to our modern Western psyche. It is the art of gardening the land with the artistry and the beauty that for so long has lain dormant in our souls.

The images we see on the landscape are but reflections in our social mirror of the way we treat ourselves and one another. As we compete and fight and live in fear, so we destroy the land; as we cooperate and coordinate and live in love, so we heal the land. We see the inner landscape of our being reflected on the outer landscape of the Earth; we see ourselves reflected in the care we take of the land. So, the question is, "How do we participate in creating our world and to what extent?" Do we create a world in a way that is environmentally compatible with human existence, or do we create an unfriendly environment that is hostile to human existence? Do we alienate ourselves from our own planet, or do we accept our responsibility as trustees of Nature's bounty and act accordingly? One thing is clear, nothing will change the effect on the collective, outer landscape until we first change the cause in our individual, inner landscapes, until we move toward conscious simplicity in both our inner and our outer lifestyles.

Acknowledgments
Zane Maser and Jean Matthews kindly read this paper and helped me to clarify both my thinking and how I expressed the ideas. I am grateful.

References

Blaschke, H. and Bumler, W. 1986. "Uber die Rolle der Bilogeozonose im Wurzelbereich von Waldbaumen." *Forstwissenschaft. Centralb.* 105:122–130.

Buber, M. 1970. *I and Thou.* Charles Scribner's Sons, New York, NY. 185 pp.

Campbell, J., with B. Moyers. 1988. *The power of myth.* Doubleday, New York, NY. 233 pp.

Chasan, D. J. 1977. *Up for grabs, inquiries into who wants what.* Madrona Publ., Inc., Seattle, WA. 133 pp.

Durrieu, G., M. Genard, and F. Lescourret. 1984. "Les micromammiferes et la symbiose mycorhizienne dans une foret de montagne." *Bull. Ecol.* 5:253–263.

Forsman, E. D., Mewlow, E. C., and Wight, H. M. 1984. "Distribution and biology of the spotted owl in Oregon," *Wildl. Monogr.* 87:1–64.

Franklin, J. F., K. Cromack, Jr., W. Denison, A. McKee, C. Maser, J. Sedell, F. Swanson, and G. Juday. 1981. "Ecological characteristics of old-growth Douglas-fir forests." USDA For. Serv. Gen. Tech. Rep. PNW-118, Pac. Northwest Forest and Range Exp. Sta., Portland, OR. 48 pp.

Harmon, M. E., J. F. Franklin, F. J. Swanson, P. Sollins, S. V. Gregory, J. D. Lattin, N. H. Anderson, S. P. Cline, N. G. Sumen, J. R. Sedell, G. W. Lienkaemper, K. Cromack, Jr., K. W. Cummins. 1986. "Ecology of coarse woody debris in temperate ecosystems." *Advances in Ecological Research.* Academic Press, New York, NY. 15:133–302.

Harris, L. D. 1984. *The fragmented forest.* Univ. Chicago Press, Chicago, IL. 211 pp.

Irwin, L. 1986. "Ecology of the spotted owl in Oregon and Washington." National Council of the Paper Industry for Air and Stream Improvement, Inc. (NCASI), *Tech. Bull.* No. 509. 129 pp.

Maser, C., R. F. Tarrant, J. M. Trappe, and J. F. Franklin (Tech. Eds.). 1988. "From the forest to the sea: A story of fallen trees." USDA For. Serv. PNW-GTS-229, Pac. Northwest Res. Sta., Portland, OR. 153 pp.

Meslow, E. C., Maser, C., and Verner, J. 1981. "Old-growth forests as wildlife habitat." *Trans. N. Amer. Wildl. Natl Resour. Conf.* 46:329–335.

Toynbee, A. 1958. *Civilization on trial and the world and the west.* Meridian Books, Inc., New York, NY. 348 pp.

Waring, R. H. and J. F. Franklin. 1979. "Evergreen coniferous forests of the Pacific Northwest." *Science* 204:1380–1386.

The End of the Lines

Norman Myers

As many scientists and conservationists now know, we are on the verge of wave upon wave of extinctions. As is less commonly recognized, these extinctions may change the course of evolution. Human societies have long had an effect on the planet's biological systems, leading to the end of many species and occasionally to the formation of new ones. But unless we act superfast, this impending spasm of extinction will surely surpass all others in prehistory, in terms of the number of species involved and the speed with which it will happen. As in the past, new life forms will arise, but not at a fraction of the rate they are going to be lost in the coming decades and centuries. We are surely losing one or more species a day right now out of the 5 million (minimum figure) on Earth. By the time ecological equilibrium is restored, at least one-quarter of all species will probably have disappeared, possibly a third, and conceivably even more. With so many plants and animals gone, there will be a fundamental shift in evolution itself, as evolutionary processes go to work on a vastly reduced pool of species and as a few new species arise to fill in the gaps.

Of course, there have always been spasms of natural extinctions. The demise of the dinosaurs and their kin, together with numerous

Natural History, vol. 94, no. 2, 1985.

marine species, during the "great dying" 65 million years ago, eliminated between one-third and one-half of all animal and plant groups (although there were rather fewer species overall in those far-back times than today). There was an even greater megaextinction, during the late Permian 230 million years ago, when possibly three-quarters of all species were lost, out of a much smaller total than exists today.

But whereas the extinctions of the Permian, and of five other major episodes in the prehistoric past, appear to have extended over several million years at least, the current catastrophe has begun in the last few decades and is likely to do much of its damage by the middle of the coming century. Particularly since World War II, tropical forests, which cover only 7 percent of the earth's surface, have been under assault in Latin America, tropical Africa, and in Asia. These forests, which support about half of all species, are being cleared for agriculture and logged at an enormous rate. A third of them are already gone. Ecuador has lost half its forests, and in Madagascar, which has a quarter of a million species, two-thirds of which occur nowhere else, 90 percent of the forests have been cut. This is the case for country after country. Seventy percent of these losses have been brought about by peasants who practice slash-and-burn agriculture and have had to roam farther and farther into the forests to grow subsistence crops. Another 15 percent of the loss is due to cattle ranching in Latin America—the so-called hamburger connection in which large fast-food outlets in the United States and Europe foster the clearance of forests to produce cheap beef. And 15 percent of the forests have been cut for lumber. If this pattern continues, it could mean the demise of 2 million species by the middle of the next century.

Two other crucial ecosystems that have suffered terribly in recent decades are wetlands and coral reefs. Coral reefs, for example, cover about 220,000 square miles of the ocean floor in the Indian Ocean and the western Pacific, among other areas, and support about 200,000 species. During the past fifty years, countries such as the Philippines, Kenya, and Tanzania have dredged reefs in their waters, and sometimes dynamited them, to make shipping lanes. Coral reefs are converted in some countries into a substitute for concrete. The number of species at stake is obviously far smaller in these ecosystems than in tropical forests, but they are no less important. Coral reefs, wetlands, and estuaries all have an exceptional abundance and diversity of species and are unusually complex in their ecological workings. Moreover, since these environments, along with tropical forests, are the preeminent "powerhouses" of evolutionary processes—meaning they throw off many more species than other environments—their loss would be profound.

Not only is the current catastrophe likely to be relatively instantaneous but it will also undoubtedly affect more life forms than ever before because more species exist now than ever before. According to David M. Raup and his colleagues at the University of Chicago, the average extinction "background rate" ranges between 2 and 4.6 *families* per million years, and the figure can rise to an average of 19.3 families per 1 million years during a period of mass extinctions. By contrast, in the next few decades we shall surely witness the demise of one-quarter of all plant families, or more than fifty families, together with many of the animal (mainly insect) families that live with them.

The average duration of a species is, roughly speaking, about 5 million years. According to Raup, there has been a crude average of 900,000 natural extinctions per 1 million years, or one natural extinction every 1¹/₉ years. The present human-caused rate of extinction is at least 400 times higher.

But just as significant as these losses will be the hiatus in evolutionary processes that they will cause. And we are being optimistic when we call it a hiatus. A more likely outcome is that many evolutionary developments that have taken place since the first flickerings of life will be suspended or even terminated.

The forces of natural selection can work only with the resource base they have. If that base is drastically reduced, the result could be a disruption that could persist far into the future. So far as we can discern from the geologic record, the "bounce-back" time may require several million years. Following the crash during the late Permian, for example, when the marine invertebrates lost about half their families, it took 20 million years before the survivors could establish even half as many families as they had lost.

In short, as M. E. Soule and Bruce A. Wilcox have written, "Death is one thing; an end to birth is something else."

Even as this megaextinction episode is taking place, some of the same disruptive processes causing it could lead to an outburst of speciation—the formation of new species. We can consider several categories of disruption. When species' populations are cut off from one another, which occurs when humans destroy swathes of habitat through the middle of a species' range, each population develops in its own way. Such a population eventually becomes distinct enough from the others that it can no longer interbreed with them, whereupon it ranks as a new species.

We can discern the results of this when we consider the great lakes of East Africa. In Lake Tanganyika there are 134 cichlid fish species that are endemic, that is, found nowhere else, while in neighboring Lake Malawi there are about 200 such species. Although the two lakes are no more than 200 miles apart (they were once one big lake), they possess not a single cichlid in common. At some point in the past, an ancestral species living as a single population began to split up into smaller populations, probably because of small changes in the shoreline that cut some fish off from others. Over the course of millions of years, the microhabitats of each of these populations led to the formation of hundreds of new species. In a small "off-shoot" of Lake Victoria, known as Lake Nabugabo, which only 4,000 years ago was separated from its parent lake by a sandstrip that is now less than two miles wide, there are six species of *Haplochromis*, five of them endemic. These have to rank among the "youngest" fish species known. Under natural conditions, then, fish can radiate, or throw off new species, in relatively short periods.

A second sort of disruption that can greatly accelerate natural evolutionary processes is the introduction of new resources or other materials into a species' environment. In Hawaii, for example, there were no bananas until humans introduced them about 1,000 years ago, after which several moth species of the genus *Hydylepta* emerged. The interesting thing about these new moths is that they feed exclusively on banana plants; other species of the genus feed on grasses, sedges, lilies, palms, and legumes.

In just the past few decades moths that display industrial

melanism have emerged in Britain; now we have mosquitoes, among other insects, that are immune to DDT and other pesticides, plus grasses that grow on the tailings of lead and copper mines. None of these are new species as yet, but they are headed in that direction. Where human exploitation of natural environments leads to fragmentation of wildlife habitats, the process of adaptation and change speeds up massively.

The disturbance of natural environments also tends to open up new niches, or ecological living spaces. This allows a few species to expand and then to diversify. In North America, for example, the spread of agriculture and the proliferation of urban refuse created a rapidly expanding niche that remained vacant until the English sparrow (sometimes known as the house sparrow) arrived in the 1850s. The sparrow exploited the man-made opportunities it found and quickly colonized the entire continent. From this sort of process, which scientists call adaptive radiation, new bird species can emerge in relatively short order. The sparrow has been in North America little more than a hundred years, but already it has developed some distinctive races, even subspecies.

The coyote appears to be another example. Because the wolf and other competitive carnivores have been widely eliminated, and because growing livestock herds offer much prey, the coyote has extended its range during the present century until it now lives in every state. As a consequence, the eastern coyote, which is about one-quarter larger than the western coyote and perhaps has some genes from the Algonquin wolf, now deserves to rank as a subspecies.

All of these are examples of the "creative" effects stemming from human disruption of the environment. But this marked acceleration of speciation will not remotely match the amount of extinction. Whereas extinction can occur in just a few decades, and sometimes in just a year or two (a valley in a tropical forest with a pocket of endemic invertebrates may be converted into pastureland within a single season), the time required to produce a new species is much longer. It takes decades for outstandingly capable contenders such as certain insects; centuries, if not millennia, for many other invertebrates; and hundreds of thousands or even millions of years for most mammals.

The most important factor in encouraging speciation is that a broad-scale extinction of species will vacate many niches. This will allow new species to emerge more rapidly than when there are diverse and abundant numbers of species, as there are today. But we know too little about this process in principle, and it will take too long to reveal itself in practice, for us to make any prognostications right now.

What we can suppose with some confidence is that the reduced stock of species that survives the present burst of extinction will likely contain a disproportionate number of opportunistic, or "clear," species. Such species rapidly exploit newly vacant niches (by making widespread use of food resources), are usually short-lived (with brief gaps between generations), feature high rates of population increase, and are adaptable to a wide range of environments—all traits that enable them to exploit new environments and to make excellent use of "boom seasons." These are the attributes that enable opportunistic species to prosper in a human-disrupted world, where they often become pests. Examples include the English sparrow, the European starling, the housefly, the rabbit, and the rat, plus many "weed" plants.

But this trend toward opportunism carries with it a sizable cost. The sparrow, for example, is adept at usurping the food stocks and nesting sites of many native North American birds, notably bluebirds, wrens, and swallows. Entire populations of these species are being displaced and supplanted, leaving them with less promising prospects of survival. The same thing is happening in other regions of the world that the sparrow has colonized with our help, especially South America, South Africa, southeastern Australia, and New Zealand.

A parallel situation has arisen in northwestern Europe. The herring gull's populations in the North Atlantic are now thirty times larger than at the beginning of the century, mainly because of the growing amounts of domestic garbage, sewage, and fishing offal. As a result, several other birds, notably the rarer terns, are suffering through competition for nesting space: in Britain, there are now only about 2,000 roseate terns left, or one-quarter as many as in the mid-1960s. Both the sparrow and the gull illustrate how the rise of opportunistic species can contribute further to the extinction pattern.

If generalist, or opportunistic, species will profit from the coming crash, the specialists, notably predators and parasites, will probably suffer disproportionately high losses. This is because their life styles are much more refined than the opportunists'; and their numbers are generally much smaller anyway. Since the specialists are often the creatures that keep down the populations of opportunists, there may be little to hold the pests in check. Today, probably less than 5 percent of all insect species deserve to be called pests. But if extinction patterns tend to favor clever species, the upshot could soon be a situation where these species increase until their natural enemies can no longer control them. In short, our descendants could find themselves living in a world with a "pest and weed" ecology.

Some observers may assert that our networks of parks and other protected areas will surely keep species alive, albeit with remnant stocks of individuals. But over the next several thousand years, even the best-protected areas will slowly lose some of their species, as habitat destruction increasingly isolates them. Several recent studies of species' distributions in parks and other islands of wildlife provide us with some insight into this process. Once established, a park is virtually certain to become an island of intact nature in a sea of human-dominated, and hence alien, environments; and the species on that island will steadily decline in number, compared with what was living there when the area was part of a "continent": an island can support only so much life. Generally speaking, if 90 percent of an original habitat is grossly disrupted, and the remaining 10 percent is protected, we can expect to save no more than about half the species in that area.

This means that, in operational terms, a tropical park of, say, 4,000 square miles (larger than Yellowstone National Park, the largest park in the conterminous United States) could well lose about half of its large mammal species in a few thousand years—unless park managers master "constructive interference," that is, management to safeguard depleted gene pools (a challenge we are barely coming to grips with). It further means that if all that eventually remains of the tropical forests of, say, Amazonia amounts to the areas currently protected, we can expect, according to Daniel Simberloff of Florida State University, to lose roughly two-thirds of the region's half million (minimum figure) species within a few thousand years.

For sure, a few thousand years is a long time by some standards. But in the conservation perspective, it is the twinkling of an evolutionary eye. The spectacular throngs of East African savannas—the vast herds of elephants, giraffes, zebras, and antelopes, together with their lions and other predators—probably face the same deadline. Because we have established so few parks of suitable size, and because our options for establishing any new parks are rapidly running out, certain observers, such as Soule and Wilcox, believe that we could well lose many, if not most, of our large mammals and birds within another 5,000 years—"large" meaning anything heavier than a few pounds. If we use the crash at the end of the Cretaceous as a model, we can anticipate that the eventual mammal survivors of the imminent crisis will likely be small-sized creatures, such as bats, rodents, rabbits, small canids, weasels, and raccoons.

More important, a park-based strategy for protecting species would not help much in safeguarding the future course of evolution. Of course, were we to establish extensive networks of outsized parks, we could keep many more species alive. But this would certainly not assure the perpetuation of evolutionary processes as they operate today. Even fairly small mammals apparently need unexpectedly large areas if they are to retain their capacity to throw off new species. To put it optimistically, we are probably foreclosing the prospect that our descendants will ever see a distinctive variation of the tiger or the cheetah or any other large cat, all of which require huge ranges to maintain the size of populations on which natural selection can work.

All this raises questions virtually as large as conservationists can visualize. Would we be content to hold on to the present pool of species, vast as that task would be? Or should we not pay equal attention to the future of evolution? Ian R. Franklin of the Genetics Resources Laboratory near Sydney, Australia, frames the issue this way: Do we wish to preserve precise phenotypes of particular species, that is, maintain the status quo, or do we wish to maintain phylogenetic lines that will permit evolutionary adaptations to persist, thereby leading to new species? Is it sufficient for us to keep, for example, just the two elephant species we already have, or should we also try to hold open the option of further elephantlike species in the distant future?

These are enormous questions, with enormous implications for conservation strategy. Elephants, along with most other mammals, are inclined to move around a good deal, which enables them to maintain gene flows across large territories. As a result, their gene pools are fairly uniform in scope. In this respect, an elephant in East Africa may not be so different from one 2,000 miles away in South Africa. Some biologists would assert that the remaining stocks of elephants, extensive as they are, may well be the minimum to keep open the possibility of speciation.

Furthermore, elephants have very slow breeding rates (one calf every fourth year at best). Many insects, by contrast, have immense breeding capacities and rapid turnover rates, offering them quick adaptability to ecological changes. Genetic changes get passed along quickly. This leaves them well suited to survive the environmental upheavals of human activities in the decades ahead. Elephants, whales, and other mammals that reproduce slowly, with a comparatively tiny capacity for genetic adaptation, will be at extreme evolutionary disadvantage. All the more, they should get special attention from us.

These, then, are some of the issues for *Homo sapiens* to bear in mind as we begin to impose a fundamental shift on evolution's course. The biggest dilemma by far is that as we proceed on our disruptive way, we give scarcely a moment's thought to what we are doing. If we thought about it a bit more, would this truly be what we want?

Unfortunately we are "deciding" without even the most superficial reflection—deciding all too unwittingly, yet effectively and increasingly. The impending upheaval in evolution's course could rank as one of the greatest biological revolutions of paleontological time. It will equal in scale and significance the development of aerobic respiration, the emergence of flowering plants, and the arrival of limbed animals. But of course, the prospective degradation of many evolutionary capacitics will be an impoverishing, not a creative, phenomenon.

In short, the future of evolution should rank as one of the most challenging problems that humankind has ever encountered. After all, we are the first species ever to be able to look upon nature's work and to decide whether we would remake part of it—to consciously determine evolution by what we do or don't do.

Fortunately there are signs—a few glimmerings—of a new spirit emerging. The mass extinction of species is no longer seen as a preoccupation of cutesy-creature enthusiasts and ecofreaks. It is being perceived as a vast rending of the fabric of life with pragmatic implications for citizens around the world, now and for generations to come. In turn, the message is being picked up by political leaders—most of all by the government of the United States. In 1981, the State Department convened an international Strategy Conference on Biological Diversity (meaning biological depletion). And in 1983 Congress passed the International Environment Protection Act, which requires the government, through its foreign aid programs, among other activities, to take special account of species communities and gene reservoirs throughout the world.

By comparison with the needs of the situation, this can all be viewed as far too little and far too late. But it is a start toward recognizing one of the great sleeper issues of our time. The first great waves of extinctions are only beginning to wash over the earth. We can still save species by the millions. Should we not consider ourselves fortunate that we alone among generations are being given the chance to support the right to life of a large share of our fellow species and to safeguard the creative capacities of evolution itself?

How Big Is Big Enough?

Joan Bird

Thirty-two years ago, two eminent biologists named MacArthur and Wilson published a book which sent ripples through the mainstream (and most of the tributaries) of ecology. *The Theory of Island Biogeography* is perhaps the most frequently cited manuscript in ecology, and has given rise to a flood of research in genetics, population dynamics, niche theory, bioenergetics, and conservation biology. In the last few years, some of those ripples have trickled out of the ivory tower and are lapping at current public and private land-management practices. Newmark's research is one of those ripples. [William D. Newmark is a scientist whose studies of the relationship between the size of areas and the numbers of species to be found in those areas led him to the conclusion that our national parks are too small to insure the long-term survival of resident mammalian species.] What relevance does this body of literature have for the inland West?

It is important to understand that this theory is not really about islands in the Robinson Crusoe sense. Island biogeography is about a phenomenon of which islands are only a good example. What island biogeography is really about is *isolation*, some discontinuity in

Northern Lights, July 1989.

the environment which effectively isolates a species, or a natural community. To a cottonball marsh pupfish in Death Valley, a pond is as much of an "island" as Isabela Island in the Galápagos is to a land tortoise. The elusive rosy finches breed only in the barren alpine tundra above timberline. In the West, three different species of rosy finch have evolved, each found in a separate group of mountains. Like the Galápagos finches, each of these species has evolved to fit its "island" in the sky.

The primary principle of island biogeography is the species/area rule, which means that the larger the area of an "island" the more different kinds of species it can support. Because it is simplistic, this rule has been widely challenged. Intuitively, one would guess that the more variable an "island" is, the more different kinds of species it will support. A flat desert island will support a limited number of species no matter how large it is. But an island of the same area with volcanic peaks, steaming jungles, mountain streams, ravines, mangrove swamps, mudflats, and all the accompanying tropical verdure will be throbbing with life. It is also intuitively obvious that the larger an area is, the more likely it is to be variable. So it is not surprising that considerable evidence supports the species/area rule.

What this has to do with the Northern Rockies and the rest of the world is its applicability to conservation. The fact that species are currently going extinct faster than they ever have in the history of the planet, *including* during the end of the age of dinosaurs, has made the species/area rule a subject for discussion by politicians and bureaucrats as well as biologists. Everyone knows that even the extinction crisis will *not* put a moratorium on development. But in making decisions about which kinds, shapes, and sizes of native landscape to preserve and which to alter, the legacy of MacArthur and Wilson is having an effect. In designing preserves of native landscape, one effect is the scientific basis for a "bigger is better" philosophy. Debate volleyed back and forth in the scientific literature for several years, and in the end an unusual event came to pass. The experts agreed. One big preserve is better than the same number of acres preserved in smaller chunks. So for those public and private land managers whose job it is to preserve examples of the original landscape, the big question is, "Is it big enough?"

In the eastern Great Plains states, cemeteries and railroad right-of-ways are some of the best remaining examples of the original landscape. These "islands" may harbor the last remaining population of a white-fringed prairie orchid, but no one is going to claim that these tiny refuges *are* the tall grass prairie. The West, in spite of massive resource development, still has opportunities to preserve *big* pieces—relatively intact ecosystems. In the thinly populated Northern Rockies, those opportunities are still outstanding.

Embracing those opportunities is not, however, the dream of the average westerner. Public land ownership is already extensive in western states, and resentment of government control is an ingrained attitude. Last year, Representative Morris Udall of Arizona introduced a bill in Congress which would have established a permanent national trust for public acquisition of significant lands. The opposition to this "American Heritage Trust Act" gathered momentum during the change in administrations and was born out of that western antipathy for government ownership of land.

While Udall may be able to override that sentiment with the political mass of the populous East, that basic western distaste for turning private land public will not be buried. New methods for preservation will have to be chiseled into the western tradition. And that is exactly what's happening. Private landowners in several western states are giving up the right to develop their own land for the sake of preserving habitats and species (and getting modest tax-breaks in the process). Along Montana's Blackfoot River corridor, these "conservation easements" now exist on 9,000 acres. Voluntary protection agreements, though nonbinding, are also effective short-term protection strategies in use in the western states.

A second principle of island biogeography is the relationship between degree of isolation, and diversity. For islands in the ocean, this is usually distance from mainland. For terrestrial "islands," it could mean distance to the next desert pond (might a young pupfish be able to swim there in a heavy rain?) or how formidable the barrier around or between "islands" (the ponds are very close but there is a Holiday Inn between them). If isolation is great, accidental arrivals are unlikely, so there had better be enough to start with if you want the species around for a while. (Land Manager's note: Unless you have friends in high places, *two of every kind* is not enough.)

Newmark contends that our National Parks are subject to the pitfalls of highly isolated islands. His point resonates with the conservation community, but is his premise true in the Northern Rockies? Have there been local extinctions of species, or "ecosystem decay"? The answer is "NO" for Yellowstone and Glacier National Parks. At least not yet. Still, island biogeography has had an effect on the management of these parks. The new wave of scientifically grounded conservation has spawned yet another term which is beginning to make its way out of scientific circles and into the language of management—*landscape ecology.* No longer does the management of national parks and nature preserves focus on drawing a line around the area and blithely ignoring activities outside the boundary. That Yellowstone is *not* an island but part of the Greater Yellowstone Ecosystem is an outgrowth of this greater contextual awareness.

The inland West is lucky, in that the effective area of its major national parks is increased by adjacent national forests. Chris Servheen, team leader for the grizzly bear recovery team, says that 60 percent of the grizzly habitat in the Yellowstone ecosystem is on national forest land, and without it there could be no grizzlies in Yellowstone. The National Forest Management Act and the Endangered Species Act require the U.S.D.A. Forest Service to protect habitat for endangered species and to "provide for diversity of plant and animal communities." While this is a relatively new role for the Forest Service and not totally integrated into its management practices, the agency appears to be accepting it about as fast as a bureaucracy of that size possibly can.

In its own regulations the Forest Service has committed itself to both managing for "viable populations of native species" and to figuring out exactly what that means. Commitment to a management philosophy that prevents extinction requires a deep involvement in the principles of island biogeography. More signs of that commitment are gradually emerging.

The Forest Service, and to a lesser extent the National Park Service, are beginning to use conservation easements on adjacent private land to increase their effective area. Combined with sound management of the public lands, easements will reduce the future possibility of local extinctions—ecosystem decay—in these protected areas.

The prognosis for maintaining representative examples of the original landscape and native flora and fauna is good for the Northern Rockies. But the job ahead is enormous. Recreational development is rapidly catching up with natural resource extraction in its degree of threat to unprotected natural communities. There may be reasons for protection which transcend current scientific arguments for species and habitat preservation.

That the Northern Rockies is one of the most pristine regions in the country seems to speak to something deep in the American character. A relationship with the wild may be embedded in the ancient brainstem of our species. Recreation is a threat, but it may also hold a promise of salvation. As Catholics make pilgrimages to Lourdes to be healed, a new kind of pilgrim is drawn to this region, not so much with physical infirmities as with psychic ones. While its numbers and remarkable adaptability leave *Homo sapiens* far from the endangered species list, even the most tunnel-visioned conservation biologist will admit that the future of this primate is uncertain. Not because of a shortage of anything as tangible as food, breeding habitat, or cover, but perhaps more from a gap between our problem-solving capacities and our problem-creating ones.

Could there be something in a natural ecosystem which humans need for their survival? In the last twenty years, state natural area programs have caught on with all the fervor of camp revivals in states where native landscapes are essentially gone. At tremendous public and private cost, agricultural lands are being "restored" to a diverse, native condition, and restoration ecology has emerged as a new scientific discipline.

In a recent article in the *Natural Areas Journal*, Gordon Whitney of the Yale Forestry School offers an argument for the preservation of old-growth forest, which is just as applicable to natural ecosystems of all kinds. He calls it "The Heritage Argument," and raises the question "What role has the natural environment played in the development of the American character?" This is not an easy question to address, but Whitney answers with quotes from Aldo Leopold:

> It fostered a certain vigorous individualism combined with ability to organize, a certain intellectual curiosity bent to practical ends, a lack of subservience to stiff social forms, and an intolerance of drones.

These qualities would be handy for tackling the current problems threatening the existence of human and the rest of life on earth. In this same article, Leopold continues:

> Is it not a bit beside the point for us to be so solicitous about preserving [America's] institutions without giving so much as a thought to preserving the environment which produced them and which may now be one of our effective means of keeping them alive?

We really don't know what all the needs are for the survival of our species. But in providing for the continued existence of the rest of this living family, we may very well be saving some nutrient essential to the human spirit, some antidote for the life-destroying behaviors that plague us.

From The Arrogance of Humanism

David Ehrenfeld

The Conservation Dilemma

Consider the lilies of the field, how they grow; they toil not, neither do they spin: And yet I say unto you, That even Solomon in all his glory was not arrayed like one of these.

—MATTHEW 6:28–29

Man is accustomed to value things to the extent that they are useful to him, and since he is disposed by temperament and situation to consider himself the crowning creation of Nature, why should he not believe that he represents also her final purpose? Why should he not grant his vanity this little fallacy? . . . Why should he not call a plant a weed, when from his point of view it really ought not to exist? He will much more readily attribute the existence of thistles hampering his work in the field

> *to the curse of an enraged benevolent spirit, or the malice of a sinister one, than simply regard them as children of universal Nature, cherished as much by her as the wheat he carefully cultivates and values so highly. Indeed, the most moderate individuals, in their own estimation philosophically resigned, cannot advance beyond the idea that everything must at least ultimately redound to the benefit of mankind, or indeed that some additional power of this or that natural organism may yet be discovered to render it useful to man, in the form of medicine or otherwise.*
>
> —JOHANN WOLFGANG VON GOETHE,
> *"An Attempt to Evolve a General Comparative Theory"*

The cult of reason and the modern version of the doctrine of final causes interact within the humanist milieu to bolster one another; one result is that those parts of the natural world that are not known to be useful to us are considered worthless unless some previously unsuspected value is discovered. Nature, in Clarence Glacken's words, is seen as "a gigantic toolshed," and this is an accurate metaphor because it implies that everything that is not a tool or a raw material is probably refuse. This attitude, nearly universal in our time, creates a terrible dilemma for the conservationist or for anyone who believes of Nature, as Goethe did, that "each of her creations has its own being, each represents a special concept, yet together they are one." The difficulty is that the humanistic world accepts the conservation of Nature only piecemeal and at a price: there must be a *logical, practical* reason for saving each and every part of the natural world that we wish to preserve. And the dilemma arises on the increasingly frequent occasions when we encounter a threatened part of Nature but can find no rational reason for keeping it.

The Arrogance of Humanism (New York: Oxford University Press, 1978).

Conservation is usually identified with the preservation of natural resources. This was certainly the meaning of conservation intended by Gifford Pinchot, founder of the national forest system in the United States, who first put the word in common use. Resources can be defined very narrowly as reserves of commodities that have an appreciable money value to people, either directly or indirectly. Since the time that Pinchot first used the word, it has been seriously overworked. A steadily increasing percentage of "conservationists" has been preoccupied with preservation of natural features—animal and plant species, communities of species, and entire ecological systems—that are *not* conventional resources, although they may not admit this.

An example of such a non-resource is an endangered amphibian species, the Houston toad, *Bufo houstonensis*. This lackluster little animal has no demonstrated or even conjectured resource value to man; other races of toad will partly replace it when it is gone, and its passing is not expected to make an impression on the *Umwelt* of the city of Houston or its suburbs. Yet someone thought enough of the Houston toad to give it a page in the International Union for the Conservation of Nature's lists of endangered animals and plants, and its safety has been advanced as a reason for preventing oil drilling in a Houston public park.

The Houston toad has not claimed the undivided attention of conservationists, or they might by now have discovered some hitherto unsuspected value inherent in it; and this is precisely the problem. Species and communities that lack an economic value or demonstrated potential value as natural resources are not easily protected in societies that have a strongly exploitative relationship with Nature. Many natural communities, probably the majority of plant and animal species, and some domesticated strains of crop plants fall into this category, at or near the *non*-resource end of a utility spectrum. Those of us in favor of their preservation are often motivated by a deeply conservative feeling of distrust of irreversible change and by a socially atypical attitude of respect for the components and structure of the natural world. These non-rational attitudes are not acceptable as a basis for conservation in Western-type societies, except in those few cases where preservation costs are minimal and there are no competing uses for the space now occupied by the non-resource. Consequently, defenders of non-resources generally have attempted to secure protection for their "useless" species or environments by means of a change of designation: a "value" is discovered, and the non-resource metamorphoses into a resource.

Perhaps the first to recognize this process was Aldo Leopold, who wrote in "The Land Ethic":

> One basic weakness in a conservation system based wholly on economic motives is that most members of the land community have no economic value. . . . When one of these non-economic categories is threatened, and if we happen to love it, we invent subterfuges to give it economic importance.

ECONOMIC VALUES FOR NON-RESOURCES

The values attributed to non-resources are diverse and sometimes rather contrived; hence the difficulty of trying to condense them into a list. In my efforts I have relied, in part, on the thoughtful analyses provided by G. A. Lieberman, J. W. Humke, and other members of the U.S. Nature Conservancy. All values listed below can be assigned a monetary value and thus become commensurable with ordinary goods and services—although in some cases it would require a good deal of ingenuity to do this. All are anthropocentric values.

1. *Recreational and esthetic values.* This is one of the most popular types of value to assign to non-resources, because although frequently quite legitimate, it is also easily fudged. Consequently, it plays an important part in cost-benefit analyses and environmental impact statements, filling in the slack on either side of the ledger, according to whatever outcome is desired. The category includes items that involve little interaction between people and environments: scenic views can be given a cash value. Less remote interactions are hiking, camping, sport hunting, and the like. Organizations such as the Sierra Club stress many of these, in part because their membership values them highly. It is no coincidence, for example, that among the Australian mammals, the large, showy, beautiful, diurnal ones, those like the big kangaroos that might be

seen on safari, are zealously protected by conservationists, and most are doing fairly well. Yet the small, inconspicuous, nocturnal marsupials, such as the long-nosed bandicoot and the narrow-footed marsupial mouse, include a distressingly large number of seriously endangered or recently exterminated species.

Rarity itself confers a kind of esthetic-economic value, as any stamp or coin collector will affirm. One of the great difficulties in conserving the small, isolated populations of the beautiful little Muhlenberg's bog turtle in the eastern United States is that as they have grown increasingly scarce the black market price paid for them by turtle fanciers has climbed into the hundreds of dollars. Some have even been stolen from zoos. Endangered falcons face a similar but more serious threat from falconers, who employ international falcon thieves to steal them from protected nests.

Some of the most determined attempts to put this recreational and esthetic category on a firm resource footing have been made by those who claim that the opportunity to enjoy Nature, at least on occasion, is a pre-requisite for sound mental and physical health. Several groups of long-term mental patients have supposedly benefited more from camping trips than from other treatments, and physiologically desirable effects have been claimed for the color green and for environments that lack the monotony of man-organized space.

2. *Undiscovered or undeveloped values.* In 1975 it was reported that the oil of the jojoba bean, *Simmondsia chinensis*, is very similar in its special physical properties to oil from the threatened sperm whale. Overnight, this desert shrub of the American Southwest was converted from the status of a minor to that of a major resource. It can safely be assumed that many other species of hitherto obscure plants and animals have great potential value as bona fide resources once this potential is discovered or developed. Plants are probably the most numerous members of this category: in addition to their possibilities as future food sources, they can also supply structural materials, fiber, and chemicals for industry and medicine. A book entitled *Drugs and Foods from Little-Known Plants* lists over 5,000 species that are locally but not widely used for food, medicine, fish poison, soap, scents, termite-resisting properties, tanning, dyestuffs, etc. The majority of these plants have never been investi-

gated systematically. It is a basic assumption of economic botany that domesticable new crops and, more importantly, undiscovered varieties and precursors of existing crops still occur in Nature or in isolated agricultural settlements, and expeditions are commonly sent to find them.

Animals have potential resource uses that parallel those of plants, but this potential is being developed at an even slower rate. The possibilities for domestication and large-scale breeding of the South American vicuña, the source of one of the finest animal fibers in the world, were only recognized after its commercial extinction in the wild had become imminent. Reports of bizarre uses of animals abound: chimpanzees and baboons have been employed as unskilled laborers in a variety of occupations, and even tapirs have allegedly been trained as beasts of burden. (Archie Carr tells the wonderful, even if apocryphal, tale in *High Jungles and Low* of the Central American who decided to use his pet tapir to carry his sugar crop to market, only to discover to his horror, en route, that tapirs prefer to cross rivers not by swimming but by walking on the bottom.) The total resource potential of insects, for example, as a source of useful chemical by-products or novel substances, has barely been explored; the shellac obtained from the lac insect, *Laccifer lacca*, is one of the few classical examples of this kind of exploitation.

Some species are potential resources indirectly, by virtue of their ecological associations. The botanist Arthur Galston has described one such case involving the water fern known as *Azolla pinnata*, which has long been cultivated in paddies along with rice by peasants in certain villages in northern Vietnam. This inedible and seemingly useless plant harbors colonies of blue-green algae in special pockets on its leaves. The algae are "nitrogen-fixing," that is they turn atmospheric nitrogen, the major component of air, into nitrogen fertilizer that plants can use, and this fertilizer dissolves in the surrounding water, nourishing both ferns and rice. Not surprisingly, villages that have been privy to the closely guarded secrets of fern cultivation have tended to produce exceptional quantities of rice.

Species whose resource possibilities are unknown cannot, of course, be singled out for protection, but most or all communities are likely to contain species with such possibilities. Thus the undeveloped resource argument has been used to support the growing movement to save "representative," self-maintaining ecosystems in all parts of the world (an "ecosystem" is a natural plant and animal community in its total physical environment of topography, rock substrate, climate, geographical latitude, etc.). Such ecosystems range from the stony and comparatively arid hills of Galilee, which still shelter the wild ancestors of wheat, oats, and barley, to the tropical forests of the world, whose timber, food, and forest product resources remain largely unknown even as they are destroyed.

3. *Ecosystem stabilization values.* This item is at the heart of a difficult controversy that has arisen over the ecological theory of conservation, a controversy based on a semi-popular scientific idea that has been well expressed by Barry Commoner:

> The amount of stress which an ecosystem can absorb before it is driven to collapse is also a result of its various interconnections and their relative speeds of response. The more complex the ecosystem, the more successfully it can resist a stress. . . . Like a net, in which each knot is connected to others by several strands, such a fabric can resist collapse better than a simple, unbranched circle of threads—which if cut anywhere breaks down as a whole. Environmental pollution is often a sign that ecological links have been cut and that the ecosystem has been artificially simplified.

I will explain a little later why the idea that natural ecosystems that have retained their original diversity are more stable than disturbed, simplified ones is controversial; but it is listed here because it has become one of the principal rationalizations for preserving nonresources, for keeping the full diversity of Nature. A more general and much less controversial formulation of this "diversity-stability" concept is discussed separately under Item 9 in this list.

One specific and less troublesome derivation of the diversity-stability hypothesis concerns monocultures—single-crop plantings—in agriculture and forestry. It has long been known that the intensive monoculture that characterizes modern farms and planted forests leads to greater ease and reduced costs of cultivation and harvesting, and increased crop yields; but this is at the expense of higher risk of epidemic disease and vulnerability to insect and other pest attack. The reasons for this can be understood partly in terms of a reduction of species diversity. This results in much closer spacing of similar crop plants, which in turn facilitates the spread of both pests and disease organisms. It also eliminates plant species that provide shelter for natural enemies of the specialized plant pests. Monocultures also create problems in ranching and fish farming, often because of the expensive inefficiency that occurs when the single species involved makes incomplete use of available food resources. I will come back to this point shortly, when I discuss African game ranching.

4. *Value as examples of survival.* Plant and animal communities, and to a lesser extent single species, can have a value as examples or models of long-term survival. J. W. Humke has observed, "Most natural systems have been working in essentially their present form for many thousands of years. On the other hand, greatly modified, man-dominated systems have not worked very reliably in the past and, in significant respects, do not do so at present." The economic value here is indirect, consisting of problems averted (money saved) by virtue of good initial design of human-dominated systems or repair of faulty ones based on features abstracted from natural systems. This viewpoint is becoming increasingly popular as disillusionment with the results of traditional planning grows. It has occurred to some to look to successful natural communities for clues concerning the organization of traits leading to persistence or survival. H. E. Wright, Jr., has stated this non-resource value in its strongest form in the concluding sentence of an interesting article on landscape development: "The survival of man may depend on what can be learned from the study of extensive natural ecosystems."

5. *Environmental baseline and monitoring values.* The fluctuation of animal or plant population sizes, the status of their organs or byproducts, or the mere presence or absence of a given species or group of species in a particular environment can be used to define normal or "baseline" environmental conditions and to determine the degree to which communities have been affected by extraordinary outside influences such as pollution or man-made habitat alteration. Biological functions such as the diversity of species in a particular location

when studied over a period of years are the best possible indicators of the meaningful effects of pollution, just as the behavior of an animal is the best single indicator of the health of its nervous and musculoskeletal systems. Species diversity is a resultant of all forces that impinge on ecosystems. It performs an automatic end-product analysis. It should also be noted that the traditional economic value of a species is of no significance in determining its usefulness as an environmental indicator—an important point if we are concerned with the metamorphosis of non-resources into resources.

With the exception of the biological monitoring of water pollution, there are few examples of the use of hitherto "worthless" species as indicators of environmental change. In the case of water pollution, the pioneering work on indicator species has been done by the freshwater biologist Ruth Patrick, who studies the aquatic communities of algae and invertebrate animals. She and her many associates have compiled lists of the kinds and numbers of organisms that one expects to find in various waters under varying conditions of naturalness.

There are a few other examples of this use of plants and animals. Lichens, the complex, inoffensive plants that encrust trees and rocks, are sensitive indicators of air pollution, especially that caused by dust and sulfur dioxide. Few lichens grow within fifty miles of a modern urban area—the forests of early colonial America were described as white because of the lichens that covered the tree trunks, but this is no more. The common lilac develops a disease called leaf roll necrosis in response to elevated levels of ozone and sulfur dioxide. The honey of honeybees reveals the extent of heavy metal pollution of the area where the bees collected nectar. And the presence of kinked or bent tails in tadpoles may be an indicator of pesticides, acid rain, or even local climatic change. All this is reminiscent of the ancient practice of examining the flight and feeding behavior of birds for auguries of the future, although we have no way of comparing the effectiveness of the results.

6. *Scientific research values.* Many creatures that are otherwise economically negligible have some unique or special characteristic that makes them extremely valuable to research scientists. Because of their relationship to humans, orangutans, chimpanzees, monkeys,

and even the lower primates fall into this category. Squids and the obscure mollusc known as the sea hare have nervous system properties that make them immensely valuable to neuroscientists. The identical quadruplet births of armadillos and the hormonal responses of the clawed toad, *Xenopus*, make them objects of special study to embryologists and endocrinologists, respectively. The odd life cycle of slime molds has endeared these fungi to biologists studying the chemistry of cell-cell interactions.

7. *Teaching values.* The teaching value of an intact ecosystem may be calculated indirectly by noting the economic value of land-use alternatives that it is allowed to displace. For example, a university administration may preserve a teaching forest on campus if the competing use is as an extra parking lot for maintenance equipment, but it may not be so disposed towards conservation if the forest land is wanted for a new administrative center. This establishes the teaching "value" of the forest to the administration.

In one case, in 1971, a U.S. federal district judge ordered the New York State National Guard to remove a landfill from the edge of the Hudson River and restore the brackish marsh that had occupied the site previously. One of the reasons he gave, although perhaps not the most important one in his opinion, was the marsh's prior use by local high school biology classes.

8. *Habitat reconstruction values.* Natural systems are far too complex for their elements and functional relationships to be fully described or recorded. Nor can we genetically reconstitute species once they have been wiped out. Consequently, if we wish to restore or rebuild an ecosystem in what was once its habitat, we need a living, unharmed ecosystem of that type to serve as both a working model and a source of living components. This is tacitly assumed by tropical forest ecologists, for example, who realize that clear-cutting of very large areas of tropical moist forests is likely to make it very difficult for the forest ever to return with anything like its original structure and species richness. In some northern temperate forests, strip-cutting, with intervening strips of forest left intact for reseeding and animal habitat, is now gaining favor in commercial timber operations. Actual cases of totally rebuilt ecosystems are still rare and will remain so: the best example is provided by the various

efforts to restore salt marshes in despoiled estuaries—this has been possible because the salt marsh is a comparatively simple community with only a few dominant plants, and because there are still plenty of salt marshes left to serve as sources of plants and animals and as models for reconstruction. In the future, if certain endangered ecosystems are recognized as being useful to us, then any remnant patches of these ecosystems will assume a special resource value.

9. *Conservative value: avoidance of irreversible change.* This is a general restatement of a basic fear underlying every other item on this list; sooner or later it turns up in all discussions about saving nonresources. It expresses the conservative belief that man-made, irreversible change in the natural order—the loss of a species or natural community—may carry a hidden and unknowable risk of serious damage to humans and their civilizations. Preserve the full range of natural diversity because we do not know the aspects of that diversity upon which our long-term survival depends. This was one of Aldo Leopold's basic ideas:

> A system of conservation based solely on economic self-interest is hopelessly lopsided. It tends to ignore, and thus eventually to eliminate, many elements in the land community that lack commercial value, but that are (as far as we know) essential to its healthy functioning.

What Leopold has done is to reject a blatantly humanistic approach in favor of a subtly humanistic one, and this failure to escape the humanistic bias has led to a weakness in his otherwise powerful argument. Leopold leaves us with no real justification for preserving those animals, plants, and habitats that, as Leopold knew, are almost certainly not essential to the "healthy functioning" of any large ecosystem. This is not a trivial category; it includes, in part, the great many species and even communities that have always been extremely rare or that have always been geographically confined to a small area. One could argue, for example, that lichens, which were once ubiquitous, might play some arcane but vital role in the long-term ecology of forests—this would be almost impossible to prove or disprove. But the same claim could not seriously be made for the furbish lousewort, a small member of the snapdragon family which has probably never been other than a rare constituent of the forests of Maine.

EXAGGERATIONS AND DISTORTIONS

The preceding list contains most if not all of the reasons that a humanistic society has contrived to justify the piecemeal conservation of things in Nature that do not, at first, appear to be worth anything to us. As such, they are all rationalizations—often truthful rationalizations to be sure, but rationalizations nonetheless. And rationalizations being what they are, they are usually readily detected by nearly everyone and tend not to be very convincing, regardless of their truthfulness. In this case they are not nearly as convincing to most people as the short-term economic arguments used to justify the preservation of "real" resources such as petroleum and timber.

In a capitalist society, any private individual or corporation who treated non-resources as if they were resources would probably go bankrupt at about the time of receiving the first medal for outstanding public service. In a socialist society, the result would be nonfulfillment of growth quotas, which can be as unpleasant as bankruptcy from a personal standpoint. People are not ready to call something a resource because of long-term considerations or statistical probabilities that it might be. For similar reasons, the majority of Western populations are content to live near nuclear power plants and to go on breathing asbestos fibers. Humanists do not like to worry about dangers that are out of sight, especially when material "comfort" is at stake.

If we examine the last item in the list, the "conservative value" of non-resources, the difficulty immediately becomes plain. The economic value in this case is remote and nebulous; it is protection from things that go bump in the night, the unknown dangers of irreversible change. Not only is the risk nebulous, but if a danger does materialize as the result of losing a non-resource, it may be impossible to prove or even detect the connection. Even in those cases where loss of a non-resource seems likely to initiate long-term undesirable changes, the argument may be too complex and technical to be widely persuasive; it may even be against popular belief.

An excellent if unintentional illustration of this last point has been given by the ecologist David Owen and independently by the public health scientist W. E. Ormerod. They have claimed that the tsetse fly that carries the cattle disease trypanosomiasis may be essential to the well-being of large parts of sub-Saharan Africa because it keeps cattle out of areas prone to overgrazing and the desert formation that follows. But the tsetse eradication programs continue unabated.

Because of the great complexity of environmental relationships and the myriad interlinkages among objects and events in Nature, it is also possible for ecologists and environmentalists to go to the opposite extreme and postulate future consequences from present events where in fact no connection or causal relationship is likely to exist. There are even those who, moving far beyond the ecologically reasonable, if humanistic, position of Leopold, assume that *everything* in Nature is essential to the survival of the natural world because evolution insures that everything is here for an important purpose or reason. R. Allen, for example, summed up in a popular scientific journal his reasons for relying strictly on resource arguments for preserving the richness of Nature: the economic climate is now such, he notes, that

> only the most severely practical arguments will prevail. Faint-hearted ecologists who fear that their favourite species *are* damned-well useless will just have to risk it. No doubt there is some redundancy in the system, but there are strong theoretical grounds for believing that most of the species on this planet are here for a better reason than that they are poor galactic map-readers.

Allen is saying that everything in Nature—including nearly all species—is highly interconnected and nearly everything has its own part to play in maintaining the natural order: consequently, nearly all species are significant, have resource value. Remove a species, even a seemingly trivial one from a resource standpoint, and we are more than likely to feel the consequences somehow, somewhere, some day. This is not a new idea—its scientific popularity dates back at least as far as the nineteenth-century writings of Charles Babbage and George P. Marsh. In the ninth chapter of his *Ninth Bridgewater Treatise*, Babbage stated that "earth, air and ocean, are the eternal witnesses of the acts we have done. . . . No motion impressed by natural causes, or by human agency, is ever obliterated." Twenty-seven years later, Marsh summed up 550 pages of examples of the ecological consequences of our interference with Nature by paraphrasing and extending the ideas of Babbage:

> There exists, not alone in the human conscience or in the omniscience of the Creator, but in external material nature, an ineffaceable, imperishable record, possibly legible even to created intelligence, of every act done, every word uttered, nay of every wish and purpose and thought conceived by mortal man, from the birth of our first parent to the final extinction of our race; so that the physical traces of our most secret sins shall last until time shall be merged in that eternity of which not science, but religion alone, assumes to take cognizance.

In a sense, of course, this is correct. There may be permanent traces of every act we do (although certainly not with enough information content left to make them legible to us in most cases). And there is an infinity of obscure connections in ecology, most of them unknowable: it has recently been discovered, for example, that on the island of Mauritius, in the Indian Ocean, the last few aged survivors of a kind of tree called *Calvaria major* are not producing any more saplings because the seeds, which the old trees still drop in abundance, must pass through the gizzard of a dodo before they can germinate. And the dodo, one of our earlier victims, became extinct in 1681.

But Marsh is implying more than this sort of thing. He is implying, as does Allen, that a sizable percentage of the lingering traces of our actions will have humanistic consequences—will affect resources. I cannot accept this. I agree with Marsh that the clearing of the Valley of the Ganges must have permanently altered the ecology of the Bay of Bengal in important ways. But have there been permanent and significant "resource" effects of the extinction, in the wild, of John Bartram's great discovery, the beautiful tree *Franklinia alatamaha*, which had almost vanished from the earth when Bartram first set eyes upon it? Or a thousand species of tiny beetles that we never knew existed before or after their probable extermination? Can we even be certain that the eastern forests of the United States suffer the loss of their passenger pigeons and chestnuts in some tangible way that affects their vitality or permanence, their value to us?

The best we can say is that any such loss *might* have dreadful consequences, and although this argument is powerful to me and to many other ecologists and conservationists, I have already shown what its deficiencies are. I am not so certain that Allen's "strong theoretical grounds" can protect the Houston toad, the cloud forests, and a vast host of other living things that deserve a chance to pay out their evolution unhindered by the enactment of our humanistic fantasies.

Thus the conservation dilemma is exposed: humanists will not normally be interested in saving any non-resource, any fragment of Nature that is not manifestly useful to humankind, and the various reasons advanced to demonstrate that these non-resources really are useful or potentially valuable are not likely to be convincing even when they are truthful and correct. When everything is called a resource, the word loses all meaning—at least in a humanist value system.

One consequence of the dilemma is that conservationists are provoked into exaggerating and distorting the humanistic "values" of non-resources. The most vexing and embarrassing example for conservationists concerns the diversity-stability issue discussed earlier. I must make clear at the outset, however, that the controversy among ecologists is not over the general need to preserve the biological richness of Nature—there is little argument about that—but over the particular theoretical reason advanced by Commoner and others that diverse ecosystems are more stable than impoverished ones (in a short-term sense), that they are best capable of resisting pollution and other undesirable, man-induced change. As the ecologist Daniel Goodman has said:

> From a practical standpoint, the diversity-stability hypothesis is not really necessary; even if the hypothesis is completely false it remains

logically possible—and, on the best available evidence, very likely—that the disruption of the patterns of evolved interaction in natural communities will have untoward, and occasionally catastrophic consequences.

To understand the origins of the controversy we must go back to a classic paper by the great Spanish ecologist Ramón Margalef. Margalef noted, as others had done previously, that as natural communities of plants and animals aged after some initial disturbance (a fire, the plowing of a field, a landslide, a volcanic eruption, etc.), the number of species in these communities tended to increase until a maximum was reached and a characteristic "climax" community appeared. This climax community was thought to last until the next disturbance, whenever that happened to be. The whole process of change is called "succession." A typical plant succession in an abandoned field in New Jersey or Pennsylvania would start with annual weeds such as foxtail grass and ragweed; this would change after one or two years to perennial weeds such as the goldenrods and asters; soon clumps of blackberries and other woody plants would appear, then "early successional" trees—red cedar and black cherry—would sprout from seeds dropped by birds. After ten or fifteen years, other trees such as red maples or oaks might have seeded in from the surrounding woods, and a half-century after that, the oak-hickory forest would gradually give way to the climax plant community of shade-loving trees: beech, sugar maple, and yellow birch.

To Margalef, this successional drive toward a climax community ("mature" ecosystem in his terminology) was one of several strong pieces of evidence that the late stages of succession are more "stable" than earlier ones. Because he also believed these late ecosystems to be more diverse in species and in the links or interactions among species, he claimed that this diversity was responsible for the greater stability of the mature ecosystems—that the stability was a consequence of the web-like structure of the more complex communities. From this kind of reasoning were derived analogies such as the one quoted above from Commoner, in which the strength of a late successional community was compared with that of a net. This hypothesis turned out to be a rallying point for conservationists who wished to justify with scientific reasons their originally emotional desire to protect the full richness of Nature, including the apparently useless majority of species. As Goodman put it, there is "a basic appeal [in] its underlying metaphor. It is the sort of thing that people like, and want, to believe."

Even as Margalef was refining his hypothesis, five lines of evidence were combining to undermine the part of it that I have described here. First, the results of many separate studies of terrestrial and aquatic ecosystems showed that diversity does not always increase with succession, particularly in the final phases. Second, it was discovered that the process of succession is not always so schematic and regular as once believed, and that the idea of a "climax" community is, like most such abstractions, only partially in accordance with what we see in Nature. Third, investigations of plant associations by the Cornell ecologist R. H. Whittaker and his colleagues tended to show that the interdependence and interactions of the species found together in mature communities had been somewhat exaggerated.

Fourth, a mathematical analysis by Robert May failed to confirm the intuitively attractive notion stated by Commoner that the greater the number of interactions, or links, the greater the stability of the system. May's mathematical models worked the other way: the more elements (species and species interactions) there were, the greater the fluctuation of the size of the "populations" in the system when a simulated external disturbance was applied. In theory, he found that the most diverse systems ought to be the most delicate; they were the ones at greatest risk of collapse following human-induced change.

Fifth, conservationists' own direct evidence supported May and contradicted the original hypothesis: the diverse, "mature" communities were almost always the first to fall apart under heavy human-imposed stress and were always the most difficult to protect. On the other hand, Margalef's own brilliant description of early colonizing species indicated that these residents of "immature" communities are usually resilient, opportunistic, genetically variable, and behaviorally adaptable, and have high reproductive rates. They are the vermin, weeds, and common game species, among others, the organisms that are most difficult to eradicate.

As May and others perceived, the diversity-stability hypothesis, *in the restricted sense described here*, was a case of inverted cause and effect. The most diverse communities were usually those that had occupied the most stable environments for the longest periods of time. They were dependent on a stable environment—not the reverse. They did not necessarily produce the kind of short-term, internal stability that Margalef had assumed to exist. The moral of this story underscores the poignancy of the conservation dilemma. In our eagerness to demonstrate a humanistic "value" for the magnificent, diverse, "mature" ecosystems of the world—the tropical rain and cloud forests, the coral reefs, the temperate zone deserts, and so on—we stressed the role they were playing in immediate stabilization of their own environments (including their own component populations) against the pollution and other disruptive by-products of modern civilization. This was a partial distortion that not only caused less attention to be paid to the real, transcendent, long-term values of these ecosystems, but also helped to obscure, for a while, their extreme fragility in the face of human "progress."

Many different kinds of "stability" are indeed dependent on maintaining biological diversity—the richness of Nature. This is especially evident today in those places, often tropical, where soils are prone to erosion, to the loss of nutrients, and to the formation of brick-like "laterite" crusts, and where desert formation can occur; but none of these effects, however deadly and durable, is ever likely to be as easy to explain to laymen as the "stable net" hypothesis.

A much less complex example of an exaggeration or distortion that has resulted from the impulse to find values for non-resources concerns African game ranching. In the 1950s and 1960s it was first pointed out that harvesting the native wild animals of the bush and savanna might produce at least as much meat per acre as cattle raising, without the destruction of vegetation that is always associated with cattle in arid environments. This suggestion cannot be faulted in terms of ecological theory, which recognizes that the dozens of species of large native herbivores—such as gazelles, wildebeest, zebras, giraffes—eat different parts of the vegetation or the same vegetation at different times, and therefore that the environment can tolerate its native grazers and browsers far better than an equivalent or even smaller number of cattle, all of which are eating the same things. Nor is there a problem concerning food tolerance: Africans are accustomed to eat and enjoy a wide variety of animals, ranging from rodents, bats, and anteaters to monkeys, turtles, snails, locusts, and flies.

The pitfalls in this straightforward plan have only recently appeared. Apart from serious cultural problems concerning the high social value of cattle in some African tribes, which makes these Africans reluctant to reduce the size of their herds, the major drawback is ecological. The early game ranching theory and the subsequent "cropping" programs of Ian Parker tacitly assume that the populations to be cropped will replace the animals that are lost, or, to put it another way, that the populations of edible wild herbivores will be able to adapt to a heavy annual loss to market hunters. This is no doubt true of some of the more fecund species, but not all are likely to reproduce quickly enough to stand the strain of this sustained mortality. The population dynamics and management ecology of nearly all species are still largely unknown, and exploitation, legal and illegal, is proceeding with little more than speculation about the long-term consequences. In a recent ecological study it was shown that massive grazing by wildebeest during their annual migrations is necessary to provide a lush mat of grasses that can be eaten by Thompson's gazelles months later. How many other such relationships are there of which we know nothing?

The issue here is the danger of assuming, with an air of infallibility, that one knows what the ecological effects of game ranching will be. This again is a manifestation of the arrogance of humanism: if the animals are to be considered resources and worthy of being saved, then they must be available for exploitation. But our ignorance of the effects of cropping has been repeatedly underscored by Hugh Lamprey and others most knowledgeable about East African ecology. In his masterful book *The Last Place on Earth*, Harold Hayes recounts these ecological arguments and beautifully illustrates many of them with an anecdote told to him by John Owen, the noted former park director at Serengeti. Owen was describing the controversy over the return of elephants, 2,000 strong, to Serengeti and the alleged damage they were doing to the park ecosystems. Should the elephants be cropped, was the question to be decided—each side had its advocates.

> When I would come down from Arusha the wardens would take me around and show me the trampled acacias. Next day the scientists [ecologists from the Serengeti Research Institute] would take me out and show me the new acacia shoots blooming in another part of the park. Acacia seeds are carried and fertilized by elephant dung.

At this time, much of the trouble is with poachers, and there is admittedly the remote possibility that supervised game ranches and cropping schemes on a large scale will have the effect of making poaching (for cash sale) uneconomical. But there is also the possibility that game ranching and cropping will affect species diversity and ecosystem stability as much as poaching or even, in some cases, cattle raising. In our haste to preserve zebra, wildebeest, dik dik, and springbok by endowing them with a tangible humanistic value, we may have exaggerated one type of resource potential (they have many others) and in the process endangered them still further.

One of the lessons of the examples cited above is that conservationists cannot trust the power assumptions and the doctrine of final causes any more than other people can—they must not assume that ecological theory can always be made to support their cases, especially when these cases concern immediate humanistic objectives and when the scope of the debate has been artificially restricted by a short-term, cost-benefit type of approach. It is a serious mistake to assume that because we are at present the most conspicuous creation of Nature, each of her other myriad creatures and workings can somehow be turned to our benefit if we find the key. As conservationists use it, this is one of the more gentle and well-meaning of the humanist deceptions, but falsehoods that spring from good intentions are still falsehoods.

Another example of a situation where ecological theories, if viewed in a restricted context, do not support conservation practices was described by the tropical ecologist Daniel Janzen:

> One possible remedy [for the year-round persistence of agricultural pests and diseases in the tropics] is unpleasant for the conservationist. The agricultural potential of many parts of the seasonally dry tropics might

> well be improved by systematic destruction of the riparian and other vegetation that is often left for livestock shade, erosion control, and conservation. It might be well to replace the spreading banyan tree with a shed. . . . Some studies even suggest that "overgrazed" pastures may have a higher overall yield than more carefully managed sites, . . . especially if the real costs of management are charged against the system.

That is, Janzen has demonstrated here that it is quite possible for ecological theory to endow non-resources with a negative value, to make them out to be economic liabilities. In this particular case, long-term ecological considerations (such as the ultimate costs of erosion, soil nutrient loss, and factors related to all the items on the list given earlier) would probably militate against the short-term ecological considerations described by Janzen. But the practical net result of any conservationist's attempt to demonstrate a resource value for natural streamside and other vegetation in the seasonally dry tropics, based on ecological theory, would be to expose the conservation position to unnecessary attack.

I want to emphasize here that the purpose of this chapter is a restricted one: to demonstrate how the ubiquitous humanist assumptions taint and damage the efforts even of those who are busy fighting the environmental consequences of modern humanism, and to identify the honest, the durable, the non-humanist reasons for saving Nature. This does not mean that I reject resource arguments when they are valid. The Amazonian rain forest, the green turtle, and many other forms of life are indeed resources; they contribute heavily to the maintenance of human well-being. The prospect of their loss is frightening to anyone with ecological knowledge, and it is not my aim to make it appear less so. But this is only one of the reasons for conservation, and it should not be applied carelessly, if only because of the likelihood of undermining its own effectiveness.

ADDITIONAL RISKS

Even when it is quite legitimate to find humanistic values for quondam non-resources, it may be risky, from a conservation viewpoint, to do so. What happens is that discovering a resource role for these once-valueless parts of Nature turns out to be a quasi-solution, and a crop of residue problems soon appears. The ecologists J. Gosselink, Eugene Odum, and their colleagues have conducted an investigation to discover the "value" of tidal marshes along the coast of the southeastern United States, which—despite its scientific elegance—can serve as an illustration of these risks.

The purpose of the project was to establish a definite monetary value for tidal marshes based on tangible resource properties. Esthetic values were therefore not considered. The properties studied included the action of tidal marshes in removing pollutants from coastal waters (a kind of tertiary sewage treatment), sport and food fish production (the marshes serve as a "nursery" for young fish), the potential for commercial aquaculture, and an assortment of other hard-to-quantify functions. The final value of *intact* marsh was calculated to be $82,940 per acre. Although the computation was a complex and speculative one which might conceivably be challenged by some ecologists, I am perfectly willing to accept it. Salt marshes are valuable.

Is calling attention to this value the best way to conserve salt marshes? If a given marsh were worth less when put to competing use than in its intact condition, the answer might be "yes," provided that the marsh were publicly owned. But discovering value can be dangerous; in effect one surrenders all right to reject the humanist assumptions.

First, any competing use with a higher value, no matter how slight the differential, would be entitled to priority in the use of the marsh site. Because most competing uses are irreversible, a subsequent relative increase in the value of marsh land would come too late. We do not generally tear down luxury high-rise apartments in order to restore tidal marshes.

Second, values change. If, for example, a new process is discovered and tertiary treatment of sewage becomes suddenly less expensive (or if the sewage acquires value as a raw material), then we will suddenly find that tidal marshes have become "worth" much less than before.

Third, the implication of the study is that both the valuable and the valueless qualities of the tidal marsh are all known and identified. Conversely, this means that those qualities of the salt marsh that have not been assigned a conventional value are not very important. This is a dangerous assumption.

Fourth, C. W. Clark has calculated that quick profits from immediate exploitation, even to the point of extinction of a resource, often are economically superior to long-term, sustained profits of the sort that might be generated by the intact resource. This economic principle has been demonstrated by the whaling industry, especially in Japan, where it has been realized that the money made from the rapid commercial extinction of whales can be reinvested in various "growth" industries, and the total profits will ultimately be greater than if the whales had been harvested at a rate that would allow them to survive indefinitely. In other words, finding a value for some part of Nature is no guarantee that it will be *rational* for us to preserve it—the reverse may hold.

Given these four objections, the risks of even legitimate reassignment of non-resources as resources become quite plain, as do the risks of over-emphasizing the humanist cost-benefit approach in conserving even the more traditional and accepted resources. There is no true protection for Nature within the humanist system—the very idea is a contradiction in terms.

There is another risk in assigning resource value to non-resources: whenever "real" values are computed it becomes possible—even necessary—to rank the various parts of Nature for the unholy task of determining a *priority* of conservation. Because dollar values of the sort worked out for tidal marshes are not often available, other ranking methods have been devised. These are meant to be applied in a mechanical, objective fashion.

One such ranking system has been developed by F. R. Gehlbach for evaluating state parkland in Texas. Properties that are scored and totaled in Gehlbach's system include "climax condition," "educational suitability," "species significance" (presence of rare, endangered, and locally unique species), "community representation" (number and type of plant and animal communities included), and "human impact" (current and potential), in order of increasing importance. Gehlbach evidently believes that the numerical scores generated by this system can be used, without additional human input, to determine conservation priorities. He states:

> It is suggested that if offered for donation [to the State of Texas], an area be accepted only when its natural area score exceeds the average scores of the same or similar community-type(s) in the natural area reserve system.

Other ranking systems exist in both Britain and the United States, and more will probably be developed.

There are two hazards of ranking the parts of Nature, and these militate against the uncritical or mechanical use of this sort of system. First there is the problem of incomplete knowledge. It is impossible to know all the properties of anything in Nature, and the more complex the entity (e.g., a natural community) the less we know. It is tempting, for example, to punch a notch in a computer card that characterizes a community as "lowland floodplain deciduous forest," and leave it at that. But such community descriptions, especially short, "objective" ones, are largely artificial abstractions; they are designed to facilitate talking about vegetation, not deciding what to do with it. It is presumptuous to assume that any formal system of ranking can serve as a substitute for personal acquaintance with the land or for human feelings—guided by information—about its meaning or value in the world of today or a hundred years from now.

The second hazard is that formal ranking is likely to set Nature against Nature in an unacceptable and totally unnecessary way. Will we one day be asked to choose between the Big Thicket of Texas and the Palo Verde Canyon on the basis of relative point totals? The need to conserve a particular community or species must be judged independently of the need to conserve anything else. Limited resources may force us to make choices against our wills, but ranking systems encourage and rationalize the making of choices. There is a difference, just as there is a difference between the scientist who finds it necessary to kill mice in order to do research, and the scientist who designs experiments in order to kill mice. Ranking systems can be useful as an adjunct to decision-making, but the more formal and generalized they become the more damage they are likely to cause.

There is only one account in Western culture of a conservation effort greater than that now taking place; it concerned endangered species. Not a single species was excluded on the basis of low priority, and by all accounts not a single species was lost.

> Of clean beasts, and of beasts that are not clean, and of fowls, and of everything that creepeth upon the earth, there went in two and two unto Noah into the ark, the male and the female, as God had commanded Noah (Genesis 7:8–9).

It is an excellent precedent.

NON-ECONOMIC VALUES

The attempt to preserve non-resources by finding economic value for them produces a double bind situation. Much of the value discovered for non-resources is indirect in the sense that it consists of avoiding costly problems that might otherwise appear if the non-resources were lost. This is the basis of the double bind. On the one hand, if the non-resource is destroyed and no disasters ensue, the conservation argument loses all capacity to inspire credence. On the other hand, if disaster does follow extinction of a supposed non-resource it may prove impossible to prove a connection between the two events.

A way to avoid this double bind is to identify the *non*-economic values inherent in all natural communities and species, and to accord them an importance at least equal to that of the indirect economic values. The first of these universal qualities might be described as the "natural art" value. It has been best articulated by the great naturalist and conservationist Archie Carr, in his book *Ulendo*:

> It would be cause for world fury if the Egyptians should quarry the pyramids, or the French should loose urchins to throw stones in the Louvre. It would be the same if the Americans dammed the Valley of the Colorado. A reverence for original landscape is one of the humanities. It was the first humanity. Reckoned in terms of human nerves and juices, there is no difference in the value of a work of art and a work of nature. There is this difference, though. . . . Any art might somehow, some day be replaced—the full symphony of the savanna landscape never.

This viewpoint is not common, and takes some getting used to, but it is apparently gaining in popularity. In an article on Brazil's endangered lion tamarins or marmosets, three species of colorful, tiny primates of the Atlantic rain forests, A. F. Coimbra-Filho advanced the notion of natural art in a frank and thoughtful statement remarkably similar to the preceding quotation:

> In purely economic terms, it really doesn't matter if three Brazilian monkeys vanish into extinction. Although they can be (and previously were) used as laboratory animals in biomedical research, other far more abundant species from other parts of South America serve equally well or better in laboratories. Lion tamarins can be effectively exhibited in zoos, but it is doubtful that the majority of zoo-goers would miss them. No, it seems that the main reason for trying to save them and other animals like them is that the disappearance of any species represents a great esthetic loss for the entire world. It can perhaps be compared to the destruction of a great work of art by a famous painter or sculptor, except that, unlike a man-made work of art, the evolution of a single species is a process that takes many millions of years and can never again be duplicated.

This natural art, unlike man-made art, has no economic worth, either directly or indirectly. No one can buy or sell it for its artistic quality, it does not always stimulate tourism, nor does ignoring it cause, for that reason, any loss of goods or services or amenities. It is distinct from the recreational and esthetic resource value described earlier and may apply to communities or species that no tourist would detour a single mile to see or to qualities that are never revealed to casual inspection.

Free as it is of some of the problems associated with resource arguments, the natural art rationale for conservation is nevertheless, in its own way, a bit contrived, and a little bit confusing. First of all, it brings up the kind of ranking problem that I discussed above. If the analogy with art holds, we would not expect all parts of Nature to have equal artistic value. Many critics would say that El Greco was a greater painter than Norman Rockwell, but is the Serengeti savanna artistically more valuable than the New Jersey Pine Barrens or the Ainsdale-Southport coastal dunes in Lancashire? And if so, what then?

Even if we concede that the art rationale for conservation does not have to foster this kind of comparison, there is still something wrong, for the natural art concept is still rooted in the same homocentric, humanistic world view that is responsible for bringing the natural world, including us, to its present condition. If the natural world is to be conserved merely because it is artistically stimulating to us, we are still conserving it for selfish reasons. There is still a condescension and superiority implied in the attitude of humans, the kindly parents, toward Nature, the beautiful problem-child. This attitude is not in harmony with the humility-inspiring discoveries of ecology or with the sort of ecological world view, emphasizing the connectedness and immense complexity of the human relationship to Nature, that now characterizes a large bloc of conservationist thought. Nor is it in accord with the growing bloc of essentially religious sentiment that approaches the same position—equality in that relationship—from a non-scientific direction.

THE NOAH PRINCIPLE

The exponents of natural art have done us a great service, being among the first to point out the unsatisfactory nature of some of the economic reasons advanced to support conservation. But something more is needed, something that is not dependent upon humanistic values. Charles S. Elton, one of the founders of ecology, has indicated another non-resource value, the ultimate reason for conservation and the only one that cannot be compromised:

> The first [reason for conservation], which is not usually put first, is really religious. There are some millions of people in the world who think that animals have a right to exist and be left alone, or at any rate that they should not be persecuted or made extinct as species. Some people will believe this even when it is quite dangerous to themselves.

This non-humanistic value of communities and species is the simplest of all to state: *they should be conserved because they exist and because this existence is itself but the present expression of a continuing historical process of immense antiquity and majesty.* Long-standing existence in Nature is deemed to carry with it the unimpeachable right to continued existence. Existence is the only criterion of the value of parts of Nature, and diminution of the number of existing things is the best measure of decrease of what we ought to value. This is, as mentioned, an ancient way of evaluating "conservability," and by rights ought to be named the "Noah Principle" after the person who was one of the first to put it into practice. For those who reject the humanistic basis of modern life, there is simply no way to tell whether one arbitrarily chosen part of Nature has more "value" than another part, so like Noah we do not bother to make the effort.

Currently, the idea of rights conferred by other-than-human existence is becoming increasingly popular (and is meeting with increasing resistance). I shall give only two examples. In a book entitled *Should Trees Have Standing?*, C. D. Stone has presented the case for existence of legal rights of forests, rivers, etc., apart from the vested interests of people associated with these natural entities. Describing the earth as "one organism, of which Mankind is a functional part," Stone extends Leopold's land ethic in a formal way, justifying such unusual lawsuits as *Byram River, et al.* v. *Village of Port Chester, New York, et al.* If a corporation can have legal rights, responsibilities, and access through its representatives to the courts ("standing"), argues Stone, why not rivers? Stone's essay has already been cited in one minority decision of the United States Supreme Court—it is not frivolous. I doubt that his suggestion will make much headway until humanism loses ground, but the weaknesses of the notion of legal standing for Nature are not important here; the mere emergence of this idea at this time is a significant event.

The ultimate example, however, of the Noah Principle in operation has been provided by Dr. Bernard Dixon in a profound little article on the case for the guarded conservation of *Variola*, the smallpox virus, an endangered species:

> Because man is the only product of evolution able to take conscious steps, whether based on logic or emotion, to influence its course, we have a responsibility to see that no other species is wiped out. . . . Some of us who might happily bid farewell to a virulent virus or bacterium may well have qualms about eradicating forever a "higher" animal—whether rat or bird or flea—that passes on such microbes to man. . . . Where, moving up the size and nastiness scale (smallpox virus, typhoid

> fever bacilli, malarial parasites, schistosomiasis worms, locusts, rats . . .), does conservation become important? There is, in fact, no logical line that can be drawn. Every one of the arguments adduced by conservationists applies to the world of vermin and pathogenic microbes just as they apply to whales, gentians, and flamingoes. Even the tiniest and most virulent virus qualifies.

In other parts of the article Dixon makes a strong case for preserving smallpox as a resource (not for biological warfare, though); nevertheless, the non-humanistic "existence value" argument is the one that matters more.

Charles Elton proposed that there were three different reasons for the conservation of natural diversity:

> because it is a right relation between man and living things, because it gives opportunities for richer experience, and because it tends to promote ecological stability—ecological resistance to invaders and to explosions in native populations.

He stated that these reasons could be harmonized and that together they might generate a "wise principle of co-existence between man and nature." Since these words were written, we have ignored this harmony of conservation rationales, shrugging off the first, or religious, reason as embarrassing or ineffective and relying on rational, humanistic, and "hard scientific" proofs of value.

I am not trying to discredit all economic and selfish uses of Nature or to recommend the abandonment of the resource rationale for conservation. Selfishness, within bounds, is necessary for the survival of any species, ourselves included. Furthermore, should we rely exclusively on non-resource motivations for conservation, we would find, given the present state of world opinion and material aspirations, that there would soon be nothing left to conserve. But we have been much too careless in our use of resource arguments—distorting and exaggerating them for short-term purposes and allowing them to confuse and dominate our long-term thinking. Resource reasons for conservation can be used if honest, but must always be presented together with the non-humanistic reasons, and it should be made clear that the latter are more important in every case. And when a community or species has no known economic worth or other value to humanity, it is as dishonest and unwise to trump up weak resource values for it as it is unnecessary to abandon the effort to conserve it. Its non-humanistic value is enough to justify its protection—but not necessarily to assure its safety in this human-obsessed world culture.

I have tried to show in this chapter the devilish intricacy and cunning of the humanists' trap. "Do you love Nature?" they ask. "Do you want to save it? Then tell us what it is good for." The only way out of this kind of trap, if there is a way, is to smash it, to reject it utterly. This is the final realism; we will come to it sooner or later—if sooner, then with less pain.

Non-humanistic arguments will carry full and deserved weight only after prevailing cultural attitudes have changed. Morally backed missionary movements, such as the humane societies, are doing quite well these days, but I have no illusions about the chance of bringing about an ethical change in our Faustian culture without prompting by some general catastrophe.

Not all problems have acceptable solutions; I feel no constraint to predict one here. On the one hand, conservationists will not succeed in a general way using only the resource approach, and they will often hurt their own cause. On the other hand, an Eltonian combination of humanist and non-humanist arguments may also fail, and if it succeeds, as Mumford has implied in "Prospect," it will probably be because of forces that the conservationists *neither expected nor controlled*:

> Often the most significant factors in determining the future are the irrationals. By "irrational" I do not mean subjective or neurotic, because from the standpoint of science any small quantity or unique occasion may be considered as an irrational, since it does not lend itself to statistical treatment and repeated observation. Under this head, we must allow, when we consider the future, for the possibility of miracles. . . . By a miracle, we mean not something outside the order of nature but something occurring so infrequently and bringing about such a radical change that one cannot include it in any statistical prediction.

But in the event of such an unexpected change in cultural attitudes, those of us who have already rejected the humanistic view of Nature will at least be ready to take advantage of favorable circumstances. And whatever the outcome, we will have had the small, private satisfaction of having been honest for a while.

The Gift of Wilderness

Wallace Stegner

1.

Once, the story goes, a squirrel could have traveled from the Green Mountains of Vermont to the swamps below Jacksonville, or from Chesapeake Bay to the Mississippi, without ever setting foot to ground. A flea on that squirrel, if he got the right transfer, could have gone on to the Staked Plains of Texas, the Uinta Basin in Utah, the upper Green River in Wyoming, or the Judith Basin in Montana, without eating anything but buffalo. If the continent's fish had decided to hold a meeting, delegates could have started from places as far apart as the West Virginia mountains and Glacier National Park, Yellowstone and the Ozarks, the Sangre de Cristos and the Minnesota height of land, and arrived unimpeded and full of health at the convention center in New Orleans.

Once, as George R. Stewart observed, "from eastern ocean to western ocean, the land stretched away without names." It is covered now with the names we have imposed on it, and the names contain our history as the seed contains the tree.

They are borrowed from the usage of a hundred Stone Age tribes,

One Way to Spell Man (Garden City, N.Y.: Doubleday, 1982).

as in Passamaquoddy, Wichita, Walla Walla. They commemorate explorers and early settlers, as in Duluth, Cooperstown, Houston. They honor Old World origins and imperial claims—New England, Virginia, Louisiana. They mark physiographic features—Detroit, Sault Sainte Marie, Rapid City—or reflect the piety of their founders—Santa Fe, St. Augustine, San Francisco. They remind us of the aspirations toward the good life and the perfected society that their namers brought from Europe—Philadelphia, Cincinnati, Communia. Sometimes a homely implement or weapon—Stirrup-Iron—or a battle or other incident—Wounded Knee, Quietus—has marked land and map forever, or so we say, with its human associations. Sometimes the name of a place, corrupted by oral transmission or misappropriated from an Indian tongue without being understood, teases us with possibilities. What shall we make of Ticklenaked Pond?

In the process of taming and naming the continent, we produced an economy that was the envy of the world and a political system that despite its clanking has been the model for individual freedom. As a civilization, we have not been so universally admired. But, good and bad, we have put ourselves on the map, and most of us have felt good about what we have done. It used to be the standard, unanimous American brag that "thriving" and "bustling" cities and "prosperous" farms now occupy what only a few years before was howling wilderness. We could cite turnpikes, canals, steamboats smoking up the Hudson and the Mississippi, immense rafts of logs coming downriver from the Wisconsin and Minnesota pineries, and could say in pride, "Look what we've done!"

Some of those cities were worth founding, and those farms feed half the world. We have made the country support, at a high level, 220 million people. But also, after nearly 500 years of "breaking" the wilderness, we have to acknowledge cut-over forests, deteriorated grasslands, eroding watersheds, decaying cities, proliferating slurbs, lakes and streams where fish can't live, air that periodically strangles us and periodically lets down acid rain. Instead of the wealth of wild creatures that once let every American feel his place in the web of life, we have remnant populations in refuges or in zoos; and some species, such as the passenger pigeons that once overwhelmed our senses with their millions, we do not have at all. We must go to Africa or the Arctic for the kind of experience that was once ours outside every frontier door. We have been fruitful, and multiplied; we have spread like ringworm from sea to sea and from the forty-ninth parallel to the Rio Grande; but in doing so we have plundered our living space. If we have loved the land fate gave us—and most of us did—we went on destroying it even while we loved it, until now we can point to many places we once pointed to in pride, and say with an appalled sense of complicity and guilt, "Look what we've done!"

But even though our environmental conscience has been made uneasy by a growing chorus of protest and warning that goes back well over a century to Thoreau, George Perkins Marsh, Muir, and John Wesley Powell, we can still be astonished by how fast it has happened, and look around us like Plains Indians wondering where the buffalo have gone. There has been some magic; they have disappeared into the ground.

The aware feel dismay, the unaware have not yet felt it. The continent has been tamed, but the average American's mind has not. Even yet there is a delusive spaciousness in our image of the continent, especially its western half where the names on the map are sparse. Free land—arable and habitable land—was pretty well gone by 1890, but the free land of the mind, the notions and assumptions bred into us by centuries of spaciousness and waste, will last a long time, and will more often be papered over than corrected. As it becomes harder to look forward to infinite promise, we either project our expectations into the new frontier of space, artificial and sterile and nonrenewable and incomprehensively expensive, or we convert the gilded future into the gilded past, warp expansive expectations into nostalgia for a golden age, sentimentalize the frontier and the frontier virtues into the grotesqueries of a Great Western Savings ad, and perpetuate our delusions with our myths.

For complex reasons, the western half of the country inherits the memory and assumes the dream. It is younger and less altered; its vast open spaces create the illusion of a continuing opportunity that its prevailing aridity prohibits. Also, much of it is federally owned, and we have grown accustomed to using it almost as freely as Americans once used the ownerless continent. Even the most protected places, national parks and wilderness areas and wild rivers, are available for many kinds of recreational and research use; and in the BLM lands and the national forests we hunt, fish, camp, hike, climb, hang-glide, and run around in dune buggies and ORV's without much hindrance or control. I am glad we can do those things, or at least some of them; but the practice indicates and perpetuates a state of mind. By universal assumption, the public lands are for public *use*. By a not altogether logical extension, the resources of the public lands—timber, grass, water, minerals—are for exploitation at cut rates. The Sagebrush Rebel who exclaimed angrily, "They've locked up that land so you can't do anything but look at it!" was speaking inaccurately, but he expressed a common point of view.

From the time when kings with tenuous imperial claims began parceling out grants to court favorites or to colonizing companies, we have operated on the notion that America is a country to be given away, or sold for a song, or appropriated by the first comer under squatters' rights. Who, during the California and later gold rushes, inquired about his right to pan a stream or dig a gravel bar or stake a claim? The gold rush was universal mass trespass that shortly created laws to legitimize itself. The mining industry has not retreated an inch from that original assumption, and a good part of the lumber industry and an important part of the cattle and sheep industries have retreated only a reluctant step or two. Contemporary squatters in the Alaska back country operate under the same unwritten law, and will be legitimized by the same processes. As Walter Webb has pointed out, only in America has the word "claim" come to mean a parcel of land.

There is another reason why the West, including Alaska, perpetuates the American dream or illusion. Americans have a centuries-old habit of dreaming westward. "Eastward I go only by force; but westward I go free," Thoreau wrote in 1862. "The future lies that way to me, and the earth seems more unexhausted and richer on that side. . . . I must walk toward Oregon, and not toward Europe." Actually, hope and the future had lain to the west for Europe, too, well before Columbus. Hence the Hesperides, the Fortunate Isles. "Going West," that World War I euphemism for dying, could not possibly have become a catch phrase with the direction reversed. Neither Europeans nor Americans can die eastward—the unknown lies the other way. Nor can they live and hope eastward, either. The grim history of the Golden Gate Bridge suggests the strength of the impulse to head west when hope is pinched out in other directions, and at the end of hope, in the face of the continent's last sunsets, to jump.

In America, it used to be said, the heavens are higher and the stars brighter; and it was easy to believe, as Thoreau did, that someday American achievements in literature, the arts, and the life of the mind would also reach higher than they had in constricted Europe. That amounted to a belief in the perfectibility and ultimately the superiority of the American character. "Else to what end does the world go on, and why was America discovered?"

It would be interesting to discuss that question with Thoreau now. The arts, literature, and the life of the mind have indeed done some flourishing in America since 1862, but it would be hard to make a case for the improvement of the American character. The land of opportunity that emancipated Americans and taught them freedom has been an enlargement for some, a trough for many. It has bred up rather more hogs than Emersons and Lincolns and Mark Twains and Ansel Adamses. It has opened the road to privilege along with the road to opportunity, and for millions, as population and wastefulness gained on resources, it has been a failed promise.

One would like to hear Thoreau on the subject of how long optimism, liberty, equality, faith in progress and perfectibility, even the indulgence of private and corporate greed, can survive the resource base that generated them. How long *does* freedom outlast riches? How long does democracy survive the shrinking of opportunity and the widening of the gap between rich and poor? What happens to the independent farmer and mechanic of our Jeffersonian illusion when the country, having wasted its bounty, begins to lose faith in itself? "America was promises," says an Archibald MacLeish poem written during the Great Depression. Indeed it was. There were largeness and hope in American lives so long as open continent stretched out ahead. (One example: Theodore Roosevelt's father founded an organization that before it was through shipped a hundred thousand homeless street children out of New York to new starts in the Middle West and beyond. Where would we send them now?) By now, the American dream even for the comparatively lucky may have shrunk to Edwin Land's ironic suggestion: an eight-hour day with two martinis at its end.

It has happened too fast for our minds to adjust to it. Thoreau believed that the woods around the Great Lakes would remain wilderness for many generations. They were leveled within forty years. Except for remnants like that in the Menominee Reservation, in Wisconsin, there are none of the old magnificent forests left in the Middle West. Or in the East, where the broad arrow of the king's navy took the first and best of the white pine, and mills and blister rust took the rest; where the chestnut is wiped out, and the elms are gone from nearly every common in New England. And what would Thoreau, an expert on lakes, make of Lake Erie, which until a few years ago was too polluted to support life, and which is only being brought back to semihealth by a concentrated, expensive struggle by citizens' groups and public agencies against the unreconstructed exponents of the American Way, the same folks who brought us the Love Canal?

Progress and perfectibility were concepts that rested easy on the American mind in 1862. They could be taken for granted as long as, in Jefferson's words, we were poor in labor and rich in land. But by 1930 or so, according to Walter Webb, the population of the United States was denser, even figuring in the open spaces of the West, than the population of Europe had been in 1500, when Europe began flowing westward toward the wide-open opportunities of the New World.

And that brings us, by a long, lugubrious detour, to the opportunities that are left us, and the choices that face us, at the beginning of the 1980s and the beginning of an administration that seems bent upon undoing all the environmental legislation of the past seventy-five years and turning us back to the damn-the-consequences practices that have left us, in all the ways of true civilization, poorer than people so naturally blessed have any right to be.

2.

For generations the machine and the garden have proposed contradictory goods to Americans. As the garden is more and more invaded, dug up, paved over, and polluted, we may, being adaptable, develop plastic lungs and stainless-steel bowels and learn to exist in the environment we have created for ourselves, as Stephen Benét's urban termites learned to sustain themselves on crumbs of steel and concrete. But it may also be that we will lose—are already losing?—touch with our humanity as we lose touch with the natural earth.

Secretary of the Interior James Watt and other believers in the machine seem to conceive of America as a warehouse packed with resources, and themselves as Hanson Loaders and Dempster Dumpsters with the duty of emptying it. Philosophers of the garden conceive America to be an intricate, interdependent organism of which man is not an irresponsible beneficiary but a living part, a participant who suffers and perhaps perishes if he mishandles too roughly the land, water, and air by which he lives, and extirpates ruthlessly the other species, plant or animal, that he thinks useless or inimical to his own welfare. Marcus Aurelius warned us about that sort of arrogance 2,000 years ago. "What is bad for the beehive cannot be good for the bee."

Obviously I am on the side of the garden. Just as obviously, I think the struggle between garden and machine will go on until the species develops either wings or horns. But I believe the dwellers in the garden will hold their own to the extent that they will save remnants of the natural world by which we can save something of ourselves. We already feel the consequences of the other course in reduced health, increasing uglification, and decreasing sanity and joy in living. We begin to feel shortages and perceive the wisdom of conserving and the madness of continued reckless raids on our earth.

Many Americans—a majority if we may believe the polls—have it in their *blood* to be members and advocates of untrammeled nature. They don't need consequences to teach them. For while we were working so ruthlessly on the wilderness, it was working on us. It altered our habits, our cuisine, our language, our expectations, our images, our heroes. It put a curve in our axe helves and a bend in our religion. It built something into our national memory; it made us a promise. Obviously that change did not happen directly to every American, and new Americans who arrived too late to be rebaptized in wildness and who know no America except the asphalt jungles may hardly have felt it at all. But it happened to enough, and enough generations, so that institutions, laws, faiths, relations with the universe were given a torque that later Americans benefited by and learned from, that laws tend to conform themselves to, that is part of a native American faith. Many people who never or rarely get to enjoy wild nature have a belief in its rightness. That is why the "mandate" upon which Mr. Watt seems to count is inevitably going to blow up in his face. For obvious reasons, the characteristic American relation with the earth persists most strongly in the West, but it has by no means vanished from New England, the South, or the Midwest; and if only as echoes of half-forgotten history, or phrases from revolutionary declarations only half understood, or in the figures of our popular heroes, most of whom were wild men—if only in these secondary and derivative ways, it is part of the American tradition to feel a bond with the wildness which Thoreau said was the salvation of the world.

The very first archetypal American in our literature was Natty Bumppo, Leatherstocking. Cooper modeled him on Daniel Boone, and he captured not only the American but the European imagination: something new under the sun, part white Indian, part noble savage, totally removed from Europe and Europe's influence, a natural philosopher who saw God in the forest and mountains and prairie; and at the same time a loner and a killer, a symbolic orphan without antecedents: that new man, the American. He was created out of loneliness and isolation and total freedom and self-reliance; he had tested himself against a thousand dangers and perfected himself in a thousand skills. He was as self-sufficient as an Indian in the woods, and more formidable. Properly speaking, he belonged to no human society, as both Indians and whites commonly did, but was the Adam of a new one. Before ever Emerson proclaimed that we had listened too long to the courtly voice of Europe, Leatherstocking had forgotten that voice—rather, he had never heard it.

At the end of *The Pioneers*, the first novel in which he appeared, in

1823, Leatherstocking is driven out of the upstate New York settlements by the increasing of people and the diminishing of game and the tightening of civilized laws, and he goes where? West, where in the last of the Leatherstocking Tales, *The Prairie*, Cooper let him die in the light of a last blazing sunset, answering the call of the remoter wilderness with a firm, confident "Here!" Cooper was an antidemocrat and a writer of flatulent prose, but until Mark Twain created Huckleberry Finn, no American writer made a more persuasive portrait of an American in all his newness. Leatherstocking was the heroic model for a whole series of mythic figures built from history: Crockett, Carson, Bridger. In spite of our contemporary black humor and our antiheroic literary theories, he influences us yet. He is there in Hemingway's code heroes; he invests such contemporary masterpieces as Faulkner's "The Bear." He gives us some vision of our possibilities as Americans; he teaches us how wildness may alter us without turning us into bloodthirsty savages. It is not civilization that clings to his buckskins, but an innate nobility and decency, and he mistrusts destructive progress as much as any Sierra Clubber.

What remains of Leatherstocking in us remains often in distorted forms. One thing that survives is an intractable independence, an impatience with law and restriction. Unfortunately, that single trait marks many kinds. Sagebrush Rebels, members of motorcycle rallies and snowmobile organizations, and juveniles who test themselves against the police express it as surely as does a climber who tests his nerve and skill against Grand Teton or El Capitan. We are a lawless nation partly because we have been a very free nation, undisciplined in the British sense. One remembers Mrs. Trollope's fretful remark that liberty in America is enjoyed by the disorderly at the expense of the orderly. Some part of Crèvecoeur's equation is still to be factored in. We still have to civilize liberty and independence, without obliterating them.

Curiously, it may be the love of wilderness that finally teaches us civilized responsibility, for wilderness, once our parent and teacher, has become our dependent.

"We are a remnant people in a remnant country," Wendell Berry has written. "We have used up the possibilities inherent in the youth of our nation; the new start in a new place with new vision and new hope. . . . We have come, or are coming fast, to the end of what we were given."

That is not quite so pessimistic a statement as it sounds. Berry himself, a Kentucky farmer-poet who farms his land with horses and laboriously restores eroded hillsides, depleted fields, and cutover woods, represents one sort of refusal to go on with the unsettling of America that he has eloquently exposed. He has made the turn that the New England Transcendentalists made long before him; he has joined nature instead of setting himself against it. And he knows as well as they did that respect for nature is indivisible. An old lady talking to her houseplants, a weekend gardener planting marigolds among his carrots and spinach, and a backpacker exultantly surveying a wilderness to whose highest point he has just won, are all on the same wavelength. In all of them, the religion of nature and the science of ecology meet. Though they may be Christians, they have left behind the Judeo-Christian tradition which puts man at the center of the universe and gives him dominion over the beasts of the field and the fowls of the air. America has taught them something besides the economics of liquidation and raid. In the same way that Indian names remain on the land, some of the Indian's reverence for the earth has become a part of us—some of us. I don't think it is an exaggeration to say that our health and even our survival as a civilization depend ultimately on how many of us learn that lesson. That means keeping alive and healthy not only our air and water but some parts of the natural continent, as many and as large parts as possible, so that nature in its wild aspects will be available to those capable of learning from it.

Once, in Hyderabad, in India, I saw a highway sign that instead of advertising some sort of goods or service made a simple statement: "Who planteth a tree is a friend of God." I believed that sign more than I believe most of ours. But I would add to it. Who preserveth a park is a friend of God. Who setteth aside a wilderness is a friend of God. Who preventeth more than the minimum of earth-moving and timber-cutting and water- and air-poisoning is a friend of God. And of man. Un-American as his motivations may seem to Joseph Coors or James Watt, he may be the best American. He may mark the beginning of the end of the long dark ages of the American success story.

3.

Preservation of wilderness came late into our priorities. As Roderick Nash showed in his fine book *Wilderness and the American Mind*, our first perceptions of the New World had as much of fear as of fascination in them. William Bradford's bleak image of the New England coast, in which, "somer being done, all things stand upon them with a wetherbeaten face; and the whole countrie, full of woods and thickets, represented a wild and savage hiew," was characteristic. The woods, like Dante's *selva oscura*, were full of terror; we should not forget that the words *wilderness* and *bewilder* are related. Shakespeare, setting his last play on a West Indian island, made its sole inhabitant a misshapen monster whose name is an anagram for cannibal. Prospero's wilderness island is made habitable only by Prospero's magic, the arts of civilization.

But as Americans familiarized themselves with the wild woods and wild beasts and wild men they had come to, what had been fearsome to all began to be home to some. William Byrd by 1729 was exulting in the freedom of the North Carolina backwoods. Crèvecoeur, by the time of the Revolution, had begun to think of the wilderness and its tribes as a sanctuary from the furies of civilized war. Cooper by 1823 was able to conceive Leatherstocking, a man compounded of natural goodness and necessary ferocity, one who could feel the woods as nature's temples almost while he lifted a scalp. Thoreau by mid-century or earlier had gone all the way to a view of wilderness as not merely fundamentally friendly but inspiring. By the mid 1860s George Perkins Marsh, in *Man and Nature*, had laid the essential foundations of the modern science of ecology, and John Muir had adopted wilderness with a passion that suggested he was substituting it for the crabbed Christianity of his father. In little more than another decade John Wesley Powell would state the principles under which the arid West should be settled. And in 1872 the nation had made its first step toward correction of a long series of wrongs against the natural continent, and set aside Yellowstone National Park.

That was the initial act of a long, developing penance that eventually saved parks, forests, and wildlife sanctuaries. But it was not until the 1930s, as a result of the vision of Aldo Leopold and Bob Marshall, that preservation of wilderness for its own sake, without reference to scenery or to recreation in the usual sense, brought the development of the public conscience to its purest, least economically motivated expression. The first wilderness areas, from the Gila onward, were preserved only by administrative action of the Forest Service, and were vulnerable to administrative reversals of policy. But in 1964, after years of effort, the Wilderness Act was finally passed, and we began the systematic program of inventorying and preserving wilderness under law.

All the way, wilderness advocates have had to battle not only robber barons and resource companies but federal bureaus as well, especially the Forest Service, dedicated to Gifford Pinchot's dubious policy of use, which in later practice has meant cheap board feet for the loggers. They have also had to resist segments of the public bent on unrestricted hunting and off-road-vehicle use in the wildest backlands. Nevertheless, the 1960s and 1970s were decades of great progress in salvage and preservation. The year 1980 saw the Alaska Lands Act, that added millions of acres to our wildlife and wilderness areas and essentially doubled our national park system. It would have been a better act, with better chances of permanence, if the Reagan victory had not made environmentalists settle for what they could get.

How much of that century and more of intelligent land-use law the Reagan administration will leave us is a question that sobers everyone with a spiritual stake in America. For the preservation of the remnants of natural America, like the conserving of natural resources instead of exploiting them at once, offers America a physically and spiritually better future than the immediate cashing of our assets would. Just as it does not seem intelligent to put the California coast at risk for a week's crude oil, so it does not seem intelligent to invade our last wilderness sanctuaries in search of a little oil, gold, molybdenum, or other minerals that we will certainly need more in the future than we need them now.

4.

Once, writing in the interests of wilderness to a government commission, I quoted a letter from Sherwood Anderson to Waldo Frank, written in the 1920s. I think it is worth quoting again. "Is it not likely," Anderson wrote, "that when the country was new and men were often alone in the fields and forest they got a sense of bigness outside themselves that has now in some way been lost . . . ? I am old enough to remember tales that strengthen my belief in a deep semireligious influence that was formerly at work among our people. . . . I can remember old fellows in my home town speaking feelingly of an evening spent on the big empty plains. It had taken the shrillness out of them. They had learned the trick of quiet."

I have a teenaged granddaughter who recently returned from a month's Outward Bound exposure to something like wilderness in Death Valley, including three days alone, with water but no food, up on a slope of the Panamints. It is a not-unheard-of kind of initiation—Christ underwent it; Indian youths on the verge of manhood traditionally went off alone to receive their visions and acquire their adult names. I don't know if my granddaughter had any visions or heard the owl cry her name. I do know *she* cried some; and I know also that before it was over it was the greatest experience of her young life. She may have greater ones later on, but she will never quite get over this one.

It will probably take more than one exposure to teach her the full trick of quiet, but she knows now where to go to learn it, and she knows the mood to go in. She has felt that bigness outside herself; she has experienced the birth of awe. And if millions of Americans have not been so lucky as she, why, all the more reason to save intact some of the places to which those who are moved to do so may go, and grow by it. It might not be a bad idea to require that wilderness initiation of all American youth, as a substitute for military service.

I, too, have been one of the lucky ones. I spent my childhood and youth in wild, unsupervised places, and was awed very early, and never recovered. I think it must have happened first when I was five years old, in 1914, the year my family moved to the remote valley of the Frenchman River, in Saskatchewan. The town was not yet born—we were among the first fifty or so people assembled to create it. Beaver and muskrat swam in the river, and ermine, mink, coyotes, lynx, bobcats, rabbits, and birds inhabited the willow breaks. During my half dozen years there, I shot the rabbits and trapped the furbearers, as other frontier boys have done, and I can remember buying Canadian Victory Bonds, World War I vintage, with the proceeds from my trapline. I packed a gun before I was nine years old. But it is not my predatory experiences that I cherish. I regret them. What I most remember is certain moments, revelations, epiphanies, in which the sensuous little savage that I then was came face to face with the universe. And blinked.

I remember a night when I was very new there, when some cowboys from the Z-X hitched a team to a bobsled and hauled a string of us on our coasting sleds out to the Swift Current hill. They built a fire on the river ice above the ford, and we dragged our sleds to the top of the hill and shot down, blind with speed and snow, and warmed ourselves a minute at the fire, and plowed up the hill for another run.

It was a night of still cold, zero or so, with a full moon—a night of pure magic. I remember finding myself alone at the top of the hill, looking down at the dark moving spots of coasters, and the red fire with black figures around it down at the bottom. It isn't a memory so much as a vision—I don't remember it, I *see* it. I see the valley, and the curving course of the river with its scratches of leafless willows and its smothered bars. I see the moon reflecting upward from a reach of wind-blown clear ice, and the white hump of the hills, and the sky like polished metal, and the moon; and behind or in front of or mixed with the moonlight, pulsing with a kind of life, the paled, washed-out green and red of the northern lights.

I stood there by myself, my hands numb, my face stiff with cold, my nose running, and I felt very small and insignificant and quelled, but at the same time exalted. Greenland's icy mountains, and myself at their center, one little spark of suffering warmth in the midst of all that inhuman clarity.

And I remember that evening spent on the big empty plains that Sherwood Anderson wrote about. In June of 1915 my father took my brother and me with him in the wagon across fifty miles of unpeopled prairie to build a house on our homestead. We were heavily loaded, the wagon was heavy and the team light, and our mare Daisy had a young foal that had a hard time keeping up. All day we plodded across nearly trackless buffalo grass in dust and heat, under siege from mosquitoes and horseflies. We lunched beside a slough where in the shallow water we ignorantly chased and captured a couple of baby mallards. Before I let mine go, I felt the thumping of that wild little heart in my hands, and that taught me something too. Night overtook us, and we camped on the trail. Five gaunt coyotes watched us eat supper, and later serenaded us. I went to sleep to their music.

Then in the night I awoke, not knowing where I was. Strangeness flowed around me; there was a current of cool air, a whispering, a loom of darkness overhead. In panic I reared up on my elbow and found that I was sleeping beside my brother under the wagon, and that a night wind was breathing across me through the spokes of the wheel. It came from unimaginably far places, across a vast emptiness, below millions of polished stars. And yet its touch was soft, intimate, and reassuring, and my panic went away at once. That wind knew me. I knew it. Every once in a while, sixty-six years after that baptism in space and night and silence, wind across grassland can smell like that to me, as secret, perfumed, and soft, and tell me who I am.

It is an opportunity I wish every American could have. Having been born lucky, I wish we could expand the opportunities I benefited from, instead of extinguishing them. I wish we could establish a maximum system of wilderness preserves and then, by a mixture of protection and education, let all Americans learn to know their incomparable heritage and their unique identity.

We are the most various people anywhere, and every segment of us has to learn all anew the lessons both of democracy and conservation. The Laotian and Vietnamese refugees who in August 1980 were discovered poaching squirrels and pigeons in San Francisco's Golden Gate Park were Americans still suffering from the shock and deprivation of a war-blasted homeland, Americans on the road of learning how to be lucky and to conserve their luck. All of us are somewhere on a long arc between ecological ignorance and environmental responsibility. What freedom means is freedom to choose. What civilization means is some sense of *how* to choose, and among what options. If we choose badly or selfishly, we have, not always intentionally, violated the contract. On the strength of the most radical political document in human history, democracy assumes that all men are created equal and that given freedom they can learn to be better masters for themselves than any king or despot could be. But until we arrive at a land ethic that unites science, religion, and human feeling, the needs of the present and the claims of the future, Americans are constantly in danger of being what Aldo Leopold in an irritable moment called them: people remodeling the Alhambra with a bulldozer, and proud of their yardage.

If we conceive development to mean something beyond earth-moving, extraction, and denudation, America is one of the world's most undeveloped nations. But by its very premises, learned in wilderness, its citizens are the only proper source of controls, and the battle between short-range and long-range goals will be fought in the minds of individual citizens. Though it is entirely proper to have government agencies—and they have to be federal—to manage the residual wild places that we set aside for recreational, scientific, and spiritual reasons, they themselves have to be under citizen surveillance, for government agencies have been known to endanger the very things they ought to protect. It was San Francisco, after all, that dammed Hetch Hetchy, it was the Forest Service that granted permits to Disney Enterprises for the resortification of Mineral King, it is Los Angeles that is bleeding the Owens Valley dry and destroying Mono Lake, it is the Air Force that wants to install the MX Missile tracks under the Utah-Nevada desert and in an ecosystem barely hospitable to man create an environment as artificial, sterile, and impermanent as a space shuttle.

We need to learn to listen to the land, hear what it says, understand what it can and can't do over the long haul; what, especially in the West, it should not be asked to do. To learn such things, we have to have access to natural wild land. As our bulldozers prepare for the sixth century of our remodeling of this Alhambra, we could look forward to a better and more rewarding national life if we learned to renounce short-term profit, and practice working for the renewable health of our earth. Instead of easing air-pollution controls in order to postpone the education of the automobile industry; instead of opening our forests to greatly increased timber cutting; instead of running our national parks to please and profit the concessionaires; instead of violating our wilderness areas by allowing oil and mineral exploration with rigs and roads and seismic detonations, we might bear in mind what those precious places are: playgrounds, schoolrooms, laboratories, yes, but above all shrines, in which we can learn to know both the natural world and ourselves, and be at least half reconciled to what we see.

PART TWO

The Impossibility of Endless Growth

The history of "civilization" has been a steady process of estrangement from nature that has increasingly developed into outright antagonism.

—Murray Bookchin

We have cherished our planet and its delicate, living biosphere much too little, and now it is telling us in many ways that it can no longer bear our onslaughts and that it is seriously ill. News comes daily of accumulating greenhouse gases, of an eroding ozone layer, of clear-cut forests and disappearing species, of poisons in meat, milk, fish, fruits, and vegetables, in the air we breathe, and in water from any source. Time *magazine, on its cover, asks, "Is anything safe?" The answer appears to be "No." Moreover, things are getting less safe at an accelerating rate, even as the growth curve for the human population is making a sharp bend toward the vertical.*

The deteriorating condition of the planet is the most important issue of our time. Growth of the human population and expansion of destructive human activities have long since become the enemies of human progress and of the biosphere in general. The "silent Armageddon" promised us decades ago by many voices in the scientific community is no longer a possibility among many possibilities lying over the horizon. It is here; we are living in its early phases.

Ecology is known as the "subversive science," because a search for the causes of environmental destruction runs headlong into our corporate-governmental system. As the extinction of species proceeds at a rate unprecedented in the life of the planet, and as exploding human populations exhaust resources and produce a deluge of wastes that poison life-support systems, the political structure continues to glorify the growth ethic and to reward short-term thinking. In a system so wed to competition the voices of conservation and holism are drowned out, and their values are crushed underfoot. Sitting Bull put it well over a century ago with his comment that our nation "is like a spring freshet that overruns its banks and destroys all who are in its path."

If there is to be a decent future, we cannot allow ourselves to remain naïve, nor can we shy away from the tasks of questioning and challenging some of the fundamental assumptions of our society. In order to rectify a bad situation it is first necessary to assess damage and to make honest attempts to understand the root causes of the problem. What we need now in much greater abundance are minds that are aware of the gravity of the situation and are willing to learn where and how constructive changes can be made.

"Apollo's Eye View," by Chellis Glendinning, is a quick aerial tour over an embattled planet: from aloft one cannot hide from the damage. Paul Shepard then asks why humans treat the planet so and presents a picture of a society that, detached from Nature, cannot mature beyond a level of childish greed. Wendell Berry, in his essay, notes little difference between modern warfare and modern industry, and he suggests that so-called industrial accidents are the revenge of Nature. Next, Van Rensselaer Potter likens uncontrolled human activity unto cancer, an analogy strengthened by the fact that the author is a world-renowned cancer researcher. Then Barry Commoner makes the case that our environmental problems cannot be solved as long as pollutants are "controlled" rather than prevented altogether.

The pieces by Edward Abbey are passionate and poetic. Abbey's grief and rage are apparent in every paragraph, and it's easy to see why his writings have served to spur radical environmentalism, and why, after his death, he continues to grow in importance.

Wayne Davis makes the connection between lifestyle and impact on the Earth when he shows us that in our wastefulness a single American does as much damage as do dozens of people collectively in a less extravagant society. Stuart Ewen then shows how advertising literally has brainwashed us into a condition of overconsumption. And finally, the Ehrlichs, Anne and Paul, people long associated with population studies, reveal that, sadly, economists and others who are guiding humanity into the future remain ignorant in areas of natural science and ecology. They get to the heart of the problem with their assertion that economics "can determine the price of everything and the value of nothing."

—BW

Apollo's Eye View: Technology and the Earth

Chellis Glendinning

In December 1972, NASA's Apollo-17 spacecraft was headed for the moon, and on that voyage a small event occurred that proved exceptional in helping modern peoples realize we live on a single, shared—and limited—planet. The first clear photograph of the whole Earth was taken.[1] Since then, this image has appeared and reappeared around the world in newspapers, books, and films and on television. It is a picture of our home, in its full, vibrant splendor.

Ancient and native peoples have long believed that this planet is a living being. They have called her Great Goddess, Earth Mother, and Gaia. But looking now beyond the beauty of her swirling cloud patterns, we detect something else. The Earth is bearing the hard, gray armor made of the steel and concrete of modern technologies. Once green with possibility, she now whirls through space in a vapor of petrochemicals and plutonium, emanating an unearthly aura of microwave, radar, and electromagnetic pulses. Look more closely.

Woman of Power, no. 11, Fall 1988.

The surface of the Earth is dotted with cities. Everywhere you see them. Enveloping the countryside with their freeways, parking lots, and skyscrapers, spouting sulfur dioxide from smokestacks and carbon monoxide from vehicles, cities are the busiest machines on the planet. Bombay, Paris, Los Angeles, Rome, Singapore—they superimpose themselves upon the Earth like mock attempts to recreate her own sacred power spots. Their extended corridors of concrete, wire, and microwave reach out to other cities until, altogether, they threaten to wrap the planet's body in one vast urban machine.

Every technology the Apollo-17 glimpses from its celestial vantage point flows from or relates back to the cities. You can see the pink lights and cooling towers of the nuclear power plants that provide energy for their operation. Generating one-twelfth of the world's electrical power,[2] reactors arise out of the land like inviolable shrines. There are 350 of them[3]—in Brokdorf, West Germany; Bilbao, Spain; Seabrook, New Hampshire; Tsuruga, Japan.

Also providing the energy required to run cities are the world's electrical power grids. From the Apollo you can glimpse high-voltage transmission lines running across the land nearly everywhere people live. In the United States alone you see over 365,000 miles of overhead lines.[4] You see thousands of power stations generating the electricity to fuel these lines—1,200 hydroelectric dams, 2,200 oil- and gas-fire plants, 900 thermal plants.[5]

There is more to see. Scanning from this photographic post in the sky, you notice vast territories blighted by the brown and gray of pollutants. See the factories on every continent burning fossil fuels, smelting metallic ores, and discharging industrial wastes. Each year human activity injects 100,000,000 metric tons of sulfur dioxide and 35,000,000 metric tons of nitrogen oxide into the atmosphere.[6] Five billion metric tons of carbon monoxide are emitted.[7] The result: whole portions of the planet's surface are wasting away. Do you see? The forests in Czechoslovakia, Poland, and West Germany are no longer hearty green but now black, withering away from the acid in the air and rain. The plants and fish in Scandinavia and eastern North America are sick and dying. Look. The loss of botanicals makes the soil less fertile. To ensure food growth, people are wielding axes, chain saws, and bulldozers to clear more trees and farm more land. They are applying more chemical fertilizers and more pesticides—causing more acidity and more dying forests.

Everywhere—in Africa and Guatemala, in South Carolina and New Zealand—chemical fertilizers are used: in 1984, 121,000,000 tons.[8] Pesticides are sprayed on farms and forests: 7,000,000,000 pounds each year.[9] Plastics are used and discarded. You can see them lying by the side of the road, leaching into groundwater in landfills, and releasing hydrogen chloride during incineration. Each day the merchant marine alone dumps 639,000 plastic containers into the sea,[10] and production of synthetic chemicals reached 320,000,000,000 pounds in 1978.[11] In fact, new chemical compositions whose effects on the body of the Earth are unknown are continually released. Fifty-five thousand human-made chemicals are now in commercial production, and each year close to 1,000 more are introduced.[12]

With all these pollutants circulating on the planet, it is no wonder her blood fouls. Look to the Danube River. She is no longer blue, but brown. The Cuyahoga River was declared "dead" in the 1970s, the Ganges is polluted, and Niagara Falls is dropping ninety metric tons of toxic chemicals over its rocks each year.[13]

Now take a look at the African continent. Whole portions are turning barren and dry. Since Western colonization—and technology—have disrupted tribal life, native populations have swelled out of proportion. And so people and governments are chopping down more trees for firewood and farmland. In eleven out of thirteen West African countries, the demand for firewood has outstripped sustainable yields.[14] With the trees gone, the soil deteriorates. It washes away with the rain and blows away in the wind. There is less evaporation from the land, and rainfall dwindles. Water tables fall. Wells dry up. Look at Ethiopia and Mauritania. Here, there is almost nothing left of life.[15]

As you scan the globe, you notice dark veins of concrete, tar, and steel crisscrossing the Earth's belly. These are the world's thoroughfares. They transport the construction materials and petrochemicals, the nuclear weapons and irradiated fuel, the commuters and vacationers of technological society. There are the railways: 1,309,137 kilometers of route intersecting the land with their tracks and coal-fired engines.[16] There are the automotive highways: in the United States alone, over 4,000,000 miles of road,[17] accommodating 348,500,000 vehicles.[18] In the Soviet automotive-manufacturing city of Togliatti, huge patches of greenery are wasting away from carbon monoxide. Mexico City's trees wither along the most traveled roadways.[19] Look to the South American continent. In Brazil the Trans-Amazon Highway plunges hundreds of miles into what a decade ago was vital rainforest. With the onslaught of civilization, great numbers of botanical species are disappearing—according to 1985 NASA photographs, 11 percent so far.[20] Note the bulldozers mowing the rainforest for development of cattle ranches. If current trends persist, a desert is predicted by the year 2050.[21]

Looking at the planet, you can also see labyrinths of pipelines bearing oil and gas. They travel offshore in the Gulf of Mexico underwater and run across the shifting deserts of the Middle East. They run from the North Sea to the British mainland, from Algeria to Italy, from Siberia to Western Europe. In the United States you can trace 227,000 miles of pipeline.[22]

Scan now for the nuclear weapons industry. Its design laboratories, production facilities, launch pads, and storage sites speckle the Earth like plates of armor. You can't see the weapons, though, despite the fact that there are some 50,000 of them.[23] They are hidden in underground silos, loaded on bombers, and circulating on trains. They are in the plains of North Dakota, central China, in the Sea of Okhotsk, under the icepack of the Arctic Ocean, and on the Mediterranean Sea.

Now, you may ask, what are those antennae protruding from the countryside? They are the military command centers—vast computers coordinating the actions of the world's arsenals. They lie below the surface of the Earth, like the underground command centers at Cheyenne Mountain in Colorado, "hardened" with tons of concrete and connected to civilian communications centers by equally "hard" cables and microwave relays.

As the Earth rotates on her axis, you notice odd structures. There's one: in the middle of the Pacific Ocean, an enormous concrete disc flattening an uncommonly treeless island. This is a nuclear dump, not to be disturbed for over 100,000 years. Look to Washington state. Here billions of gallons of radioactive wastes are hidden below ground in steel tanks, unlined trenches, and ordinary ponds.[24] Such tombs can be detected in New Mexico, Siberia, France, and India—wherever nuclear facilities operate.

There are also other waste dumps festering like carbuncles beneath the Earth's skin. In the 1970s, 35,000,000 tons of toxic wastes were generated each year in the United States,[25] while 1980 saw a leap to 63,000,000 tons—enough to fill 3,000 Love Canals.[26] The Environmental Protection Agency tells us that as many as 50,000 waste dumps exist in the United States.[27]

Sobered, you notice another strange phenomenon. Far from the cities, you notice the Earth rumbling and shaking. Then a huge plume of dust arises with no visible cause. These are underground nuclear explosions. Look to Siberia, the western United States, the South Pacific, the Sahara Desert. Like lethal assaults against her life, they gut the Earth's insides and vent poisons into her atmosphere, soil, and groundwater. There have been 1,250 such eruptions since 1945,[28] over 250 directly into the air and nearly 1,000 underground.[29]

Mining technologies also penetrate the planet. To provide the staples of technological society, they extract heavy metals, radioactive minerals, coal, oil, and asbestos as if excavating the very bowels of her body. In Bolivia, South Africa, and Zambia, you see drills invading her interior, bulldozers and dragline cranes scraping and pummeling her skin. You see the conveyor machines and processing plants readying the materials for human use. You see the coal slag heaps, uranium tailings, and stripped land left behind when the miners depart.

Look now at the seas, majestic and blue. Covering 70 percent of the Earth's surface, they too are barraged by technology. Supertankers with a capacity of 3,000,000 barrels of oil ply the oceans.[30] There are cryogenic ships transporting Liquid Natural Gas, the most fortified nonmilitary vessels in the world. There are offshore platforms and rigs along the coastlines. There are drill ships, surveyor ships, tug boats, submersible vehicles, and support ships to carry fuel and pipes to the platforms. You see freighters transporting consumer goods, resources, and irradiated fuel. You see yachts and ocean liners carrying vacationers across the seas.

Fishing is no longer an activity for boys toting bamboo poles or men with hemp nets. It is now conducted with gigantic fish-processing factories floating beside fleets equipped with sonar tracking stations, radar, computers, and high-tech ship-to-shore communications. All of these technologies increase the efficiency of fishing so that from 1950 to 1970, the worldwide catch tripled. It now exceeds the fish's capacity to reproduce.[31]

War hardware pervades the seas. At any moment, thirty-five missile-carrying submarines are diving and surfacing in the oceans. Together, they carry a total of 3,100 nuclear warheads.[32] See the iron warships cruising the Persian Gulf and the aircraft carriers in the Sea of Japan. Note the hydrophones bobbing with the waves, collecting acoustic information for submarines. See the laser and infrared detectors and the surveillance processing facilities in the water all over the Earth.

There are also technologically produced poisons in the seas. You can see them. Relatively clean waters flow into the Mediterranean at Gibraltar and quickly become darkened by pollutants from the Adriatic and Aegean Seas, from the Rhone and Nile Rivers. The Red Sea and Hiroshima Bay are dying. Each year tanker spills, automobile emissions, and polluted rivers dump 10,000,000 tons of oil into the oceans.[33] The United Kingdom's reprocessing plant at Sellafield pumps 1,200,000 gallons of radioactive wastes into the sea each day.[34]

Every living being has an aura. It is that luminous field of energy surrounding the body, and its clarity reflects the organism's well-being. The Earth's aura is cluttered with metal objects. See them floating and darting above mountain peak and valley. They are airplanes, helicopters, satellites, missiles. See the rockets arching into the sky. Five hundred are launched each year. In one day 1,700 airplanes enter and leave U.S. airspace.[35] Then there are the space stations and spacecraft—all backed up on Earth by launch sites, communications terminals, tracking installations, and data processing centers. Between 1958 and 1976, the United States, the Soviet Union, France, Britain, China, and NATO launched a total of 2,311 military and nonmilitary satellites into the atmosphere.[36] Some, like reconnaissance satellites, fly 90 to 300 miles above the Earth. Others, like communications vehicles, circle at up to 22,000 miles, each able to transmit 30,000 telephone calls and three television channels at once.[37] Together with all the lost wrenches, frozen human wastes, and spent payloads, there are some 15,000 human-made objects whirling in planetary orbit.[38]

But the Earth's aura is not only cluttered by metal objects. It is also dirty. You can see the pink and brown clouds encircling the planet from coal-gasification plants and automotive exhaust. What you can't detect are the microwave and radar rays flooding the atmosphere from communications centers, radio towers, and satellites. You can't see the electromagnetic fields perpetrated by the thousands of miles of power lines. Nor can you see the 150,000,000 tons of radioactive materials released into the Earth's stratosphere by nuclear manufacture and testing.[39] But they're there. The very photograph of the Earth that inspires this exploration, ironically, was taken by remote sensor technologies beaming microwaves at the planet.

Perhaps you've seen enough. You understand that our planet is girdled at every curve by human technologies. But there is more. If you sit long enough in our photographic mooring in the sky, you see unearthly happenings. Look here. A vast explosion rocks the Ural Mountains and leaves thousands of acres radioactive. Coal smog envelopes London, England and Donora, Pennsylvania. Mexico City is lost from sight. A bomber drops a nuclear weapon on the Spanish countryside. A satellite crashes in northern Canada. A tank of jet fuel explodes in Japan. Five thousand gallons of radioactive wastes leak into the Earth. A rocket explodes over Cape Canaveral. A cloud of methyl isolcynanate escapes from a plant in India. A reactor in the Soviet Union spews radiation around the globe.

Come back to Earth.

References

1. The first actual photograph of the whole Earth, a very crude picture, was taken aboard the U.S. satellite Explorer VI in August, 1959.
2. Amory Lovins and Hunter Lovins, *Brittle Power: Energy Strategy for National Security* (Andover, MA: Brick House Publishing, 1982), p. 141.

3. Lester Brown et al., *State of the World 1986* (New York: W. W. Norton, 1986), p. 119.
4. Economic Regulatory Admin, *The National Electricity Reliability Study*, February draft, Vol. I, DOE/RG-0055. Office of Utility System, 1981, p. 2–9.
5. Lovins, *Brittle Power*, p. 124.
6. Gene Likens et al., "Acid Rain," *Scientific American*, October 1979, p. 49; John Holum, *Topics and Terms in Environmental Problems* (New York: John Wiley and Sons, 1977), p. 619.
7. Lester Brown et al., *State of the World 1985* (New York: W. W. Norton, 1985), p. 15.
8. Lester Brown, *State of the World 1985*, p. 29.
9. Robert Wasserstrom and Richard Wiles, *Field Duty: U.S. Farmworkers and Pesticide Safety*, Study 3 (Washington, D.C.: World Resources Institute, 1985), p. 2.
10. Karin Winegar, "Plastics: The Once and Future Trash," *Utne Reader*, November–December 1987, p. 24.
11. The figures from the International Trade Commission, quoted in Ralph Nader, Ronald Brownstein, and John Richard, *Who's Poisoning America?* (San Francisco: Sierra Club Books, 1981), p. 5.
12. L. Fishbein, *Potential Industrial Carcinogens and Mutagens*, 560/5-77-005 (Washington, D.C.: U.S. Environmental Protection Agency, 1977). Thomas Maugh II, "Chemical Carcinogens: The Scientific Basis for Regulation," *Science*, September 29, 1978, pp. 1200–1205.
13. Division of Water, "Comparison of 1981–82 and 1985–86 Toxic Substance Discharges to the Niagara River," New York State: Department of Environmental Conservation, August 1987, p. 3.
14. Forestry Division, *Tropical Forest Resources* Forestry Paper 30. Rome, Italy: United Nations Food and Agriculture Organization, 1982; Gunter Schramm and David Schirah, "Sub-Saharan Africa Policy Paper: Energy," Washington, D.C.: World Bank, August 20, 1984.
15. Sara Pacher, "The World According to Lester Brown," *Utne Reader*, September/October 1987, p. 87.
16. Conversation with Louis Thompson, Railroad Division, World Bank, Washington, D.C., September 3, 1987.
17. *Highway Statistics 1985* (Washington, D.C.: Federal Highways Administration, 1985), p. 109.
18. *Highway Statistics 1985*, p. 17.
19. Lester Brown, *State of the World 1985*, pp. 106–107.

20. Randolph Harrison, "March Out of Poverty Lays Waste to Lands," *Orlando Sentinel*, March 10, 1987.
21. Frederick Ordway, *Pictorial Guide to the Planet Earth* (New York: Thomas Crowell, 1975), p. 39.
22. Testimony of Paul Tierrey, Trans Alaskan Pipeline System Hearings, Federal Energy Regulatory Commission, Washington, D.C., October 30, 1978.
23. William Arkin and Richard Fieldhouse, *Nuclear Battlefields: Global Links in the Arms Race* (Cambridge, MA: Ballinger/Harper & Row, 1985), p. 2.
24. Fred Shapiro, "Radwaste in the Indians' Backyards," *Nation*, May 7, 1983, p. 574.
25. Michael Brown, *Laying Waste: The Poisoning of America by Toxic Chemicals* (New York: Pantheon, 1979), p. 293.
26. Office of Solid Wastes, *Everybody's Problem: Hazardous Wastes* (Washington, D.C.: Environmental Protection Agency, 1981), p. 1.
27. Consultant Report to the Office of Solid Waste, *Preliminary Assessment of Cleanup Costs for National Hazardous Waste Problems* (Washington, D.C.: Environmental Protection Agency, 1979), p. 24.
28. William Arkin and Richard Fieldhouse, *Nuclear Battlefields*, p. 34.
29. "A Primer on Nuclear Testing," *Testing News*, Vol. III, No. IV, Aug. 1985.
30. Ordway, *Pictorial Guide*, p. 29.
31. Lester Brown, *The Twenty-ninth Day.* (NY: W. W. Norton, 1978), p. 22.
32. William Arkin and Richard Fieldhouse, *Nuclear Battlefields*, p. 44.
33. William Ordway, *Pictorial Guide*, pp. 132–133.
34. Richard Hudson, "Atomic Age Dump: A British Nuclear Plant Recycles Much Waste," *Wall Street Journal*, April 11, 1984.
35. William Arkin and Richard Fieldhouse, *Nuclear Battlefields*, p. 2.
36. Stockholm International Peace Research Institute (SIPR), *Outer Space: Battlefield of the Future?* (London: Taylor and Francis, Ltd., 1978), p. v, 2.
37. Peter Coy, "Satellites Affect Face of Communications," *Albuquerque Journal*, July 12, 1987.
38. Gar Smith, "Space As a Wilderness," *Earth Island Journal*, Winter 1987, p. 24.
39. Bill Lawren, "Six Scientists Who May Save the World," *Omni*, September 1987, p. 96.

From Nature and Madness

Paul Shepard

Introduction

My question is: Why do men persist in destroying their habitat? I have, at different times, believed the answer was a lack of information, faulty technique, or insensibility. Certainly intuitions of the interdependence of all life are an ancient wisdom, perhaps as old as thought itself, occasionally rediscovered, as it has been by the science of ecology in our own society. At mid-twentieth century there was a widely shared feeling that we only needed to bring businessmen, cab drivers, housewives, and politicians together with the right mix of oceanographers, soils experts, or foresters in order to set things right.

In time, even with the attention of the media and a windfall of synthesizers, popularizers, gurus of ecophilosophy, and other champions of ecology, in spite of some new laws and indications that environmentalism is taking its place as a new turtle on the political

Nature and Madness (San Francisco: Sierra Club Books, 1982).

log, nothing much has changed. Either I and the other "pessimists" and "doomsayers" were wrong about the human need for other species, and the decline of the planet as a life-support system, or our species is intent on suicide—or there is something we overlooked.

Such a something could be simply greed. Maybe the whole world is just acting out the same impulse that brought a cattlemen's meeting in West Texas to an end in 1898 with the following unanimous declaration: "*Resolved*, that none of us know, or care to know, anything about grasses, native or otherwise, outside the fact that for the present there are lots of them, the best on record, and we are after getting the most out of them while they last."

But it is hard to be content with the theory that people are bad and will always do the worst. Given the present climate of education, knowing something about grasses may *be* the greedy course if it means the way to continued prosperity.

The stockmen's resolution seems to say that somebody at the meeting had been talking new-fangled ideas of range management. Conservation in the view of Theodore Roosevelt's generation was largely a matter of getting the right techniques and programs. By Aldo Leopold's time, half a century later, the perspective had begun to change. The attrition of the green world was felt to be due as much to belief as to craft. Naturalists talking to agronomists were only foreground figures in a world where attitudes, values, philosophies, and the arts—the whole weltanschauung of peoples and nations—could be seen as a vast system within which nature was abused or honored. But today the conviction with which that idea caught the imagination seems to fade; technology promises still greater mastery of nature, and the inherent conservatism of ecology seems only to restrain productivity as much of the world becomes poorer and hungrier.

The realization that human institutions express at least an implicit philosophy of nature does not always lead them to broaden their doctrines. Just as often it backs them into a defense of those doctrines. In the midst of these new concerns and reaffirmations of the status quo, the distance between earth and philosophy seems as great as ever. We know, for example, that the massive removal of the great Old World primeval forests from Spain and Italy to Scandinavia a thousand years ago was repeated in North America in the past century and proceeds today in the Amazon basin, Malaysia, and the Himalayan frontier. Much of the soil of interior China and the uplands of the Ganges, Euphrates, and Mississippi rivers has been swept into their deltas, while the world population of man and his energy demands have doubled several times over. The number of animal species we have exterminated is now in the hundreds. An uncanny something seems to block the corrective will, not simply private cupidity or political inertia. Could it be an inadequate philosophy or value system? The idea that the destruction of whales is the logical outcome of Francis Bacon's dictum that nature should serve man or René Descartes' insistence that animals feel no pain since they have no souls seems too easy and too academic. The meticulous analysis of these philosophies and the discovery that they articulate an ethos beg the question. Similarly, technology does not simply act out scientific theory or daily life flesh out ideas of progress, biblical dogma, or Renaissance humanism. A history of ideas is not enough to explain human behavior.

Our species once did (and in some small groups still does) live in stable harmony with the natural environment. That was not because men were incapable of changing their environment or lacked acumen; it was not simply on account of a holistic or reverent attitude, but for some more enveloping and deeper reason still. The change began between five and ten thousand years ago and became more destructive and less accountable with the progress of civilization. The economic and material demands of growing villages and towns are, I believe, not causes but results of this change. In concert with advancing knowledge and human organization it wrenched the ancient social machinery that had limited human births. It fostered a new sense of human mastery and the extirpation of nonhuman life. In hindsight this change has been explained in terms of necessity or as the decline of ancient gods. But more likely it was irrational (though not unlogical) and unconscious, a kind of failure in some fundamental dimension of human existence, an irrationality beyond mistakenness, a kind of madness.

* * *

The idea of a sick society is not new. Bernard Frank, Karl Menninger, and Erich Fromm are among those who have addressed it. Sigmund Freud asks, "If the development of civilization has such a far-reaching similarity to the development of the individual and if it employs the same methods, may we not be justified in reaching the diagnosis that, under the influence of cultural urges, some civilizations—or some epochs of civilization—possibly the whole of mankind—have become neurotic?" Australian anthropologist Derek Freeman observes that the doctrine of cultural relativism, which has dominated modern thought, may have blinded us to the deviant behavior of whole societies by denying normative standards for mental health.

In his book *In Bluebeard's Castle*, George Steiner asks why so many men have killed other men in the past two centuries (the estimate is something like 160 million deaths). He notes that, for some reason, periods of peace in Europe were felt to be stifling. Peace was a lassitude, he says, periodically broken by war, as though pressures built up that had little to do with the games of national power or conflicting ideologies. He concludes that one of those pressures found its expression in the Holocaust, motivated by unconscious resentment of the intolerable emotional and intellectual burden of monotheism. Acting as the frenzied agents for a kind of fury in the whole of Christendom, the Germans sought to destroy the living representatives of those who had centuries ago wounded the mythic view of creation, stripping the earth of divine being and numinous presences, and substituting a remote, invisible, unknowable, demanding, vengeful, arbitrary god.

Steiner's approach to these seizures of extermination as collective personality disintegration has something to offer the question of the destruction of nature. What, in Steiner's framework, is indicated by the heedless occupancy of all earth habitats; the physical and chemical abuse of the soil, air, and water; the extinction and displacement of wild plants and animals; the overcutting and overgrazing of forest and grasslands; the expansion of human numbers at the expense of the biotic health of the world, turning everything into something man-made and man-used?

To invoke psychopathology is to address infancy, as most mental problems have their roots in our first years of life and their symptoms are defined in terms of immaturity. The mentally ill typically have infantile motives and act on perceptions and states of mind that caricature those of early life. Among their symptoms are destructive behaviors that are the means by which the individual comes to terms with private demons that would otherwise overwhelm him. To argue with the logic with which he defends his behavior is to threaten those very acts of defense that stand between him and a frightful chasm.

Most of us fail to become as mature as we might. In that respect there is a continuum from simple deprivations to traumatic shocks, many of which are fears and fantasies of a kind that normally haunt anxious infants and then diminish. Such primary fantasies and impulses are the stuff of the unconscious of us all. They typically remain submerged or their energy is transmuted, checked, sublimated, or subordinated to reality. Not all are terrifying: there are chimeras of power and unity and erotic satisfaction as well as shadows that plague us at abyssal levels with disorder and fear, all sending their images and symbols into dreams and, in the troubled soul, into consciousness. It is not clear whether they all play some constructive part in the long trail toward sane maturity or whether they are just flickering specters lurking beside that path, waiting for our wits to stumble. Either way, the correlation between mental unhealth and regression to particular stages of early mental life has been confirmed thousands of times over.

The passage of human development is surprisingly long and complicated. The whole of growth through the first twenty years (including physical growth) is our ontogenesis, our "coming into being," or *ontogeny.* Dovetailed with fetal life at one end and adult phases at the other, ontogeny is as surprising as anything in human biology. Anyone who thinks the human creature is not a specialized animal should spend a few hours with the thirty-odd volumes of *The Psychoanalytic Study of the Child* or the issues of the *Journal of Child Development*. In the realm of nature, human ontogeny is a regular giraffe's neck of unlikely extension, vulnerability, internal engineering, and the prospect of an extraordinary view from the top.

Among those relict tribal peoples who seem to live at peace with their world, who feel themselves to be guests rather than masters, the ontogeny of the individual has some characteristic features. I conjecture that their ontogeny is more normal than ours (for which I will be seen as sentimental and romantic) and that it may be considered to be a standard from which we have deviated. Theirs is the way of life to which our ontogeny was fitted by natural selection, fostering a calendar of mental growth, cooperation, leadership, and the study of a mysterious and beautiful world where the clues to the meaning of life were embodied in natural things, where everyday life was inextricable from spiritual significance and encounter, and where the members of the group celebrated individual stages and passages as ritual participation in the first creation.

This seed of normal ontogeny is present in all of us. It triggers vague expectations that parents and society will respond to our hunger. The newborn infant, for example, needs almost continuous association with one particular mother who sings and talks to it, breast-feeds it, holds and massages it, wants and enjoys it. For the infant as person-to-be, the shape of all otherness grows out of that maternal relationship. Yet, the setting of that relationship was, in the evolution of humankind, a surround of living plants, rich in texture, smell, and motion. The unfiltered, unpolluted air, the flicker of wild birds, real sunshine and rain, mud to be tasted and tree bark to grasp, the sounds of wind and water, the calls of animals and insects as well as human voices—all these are not vague and pleasant amenities for the infant, but the stuff out of which its second grounding, even while in its mother's arms, has begun. The outdoors is also in some sense another inside, a kind of enlivenment of that fetal landscape which is not so constant as once supposed. The surroundings are also that-which-will-be-swallowed, internalized, incorporated as the self.

The child begins to babble and then to speak according to his own timing, with the cooperation of adults who are themselves acting upon the deep wisdom of a stage of life. At first it is a matter of rote and imitation, a naming of things whose distinctive differences are unambiguous. Nature is a lexicon where, at first, words have the solid reality of things.

In this bright new world there are as yet few mythical beasts, but real creatures to watch and to mimic in play. Animals have a magnetic affinity for the child, for each in its way seems to embody some impulse, reaction, or movement that is "like me." In the playful, controlled enactment of them comes a gradual mastery of the personal inner zoology of fears, joys, and relationships. In stories told, their forms spring to life in the mind, re-presented in consciousness, training the capacity to imagine. The play space—trees, shrubs, paths, hidings, climbings—is a visible, structured entity, another prototype of relationships that hold. It is the primordial terrain in which games of imitating adults lay another groundwork for a dependable world. They prefigure a household, so that, for these children of mobile hunter-gatherers, no house is necessary to structure and symbolize social status. Individual trees and rocks that were also known to parents and grandparents are enduring counterplayers having transcendent meanings later in life.

In such a world there is no wildness, as there is no tameness. Human power over nature is largely the exercise of handcraft. Insofar as the natural world poetically signifies human society, it signals that there is no great power over other men except as the skills of leadership are hewn by example and persuasion. The otherness of nature takes fabulous forms of incorporation, influence, conciliation, and compromise. When the juvenile goes out with adults to seek a hidden root or to stalk an antelope he sees the unlimited possibilities of affiliation, for success is understood to depend on the readiness of the prey or tuber as much as the skill of the forager.

But the child does not yet philosophize on this; for him the world is simply what it seems; he is shielded from speculation and abstraction by his own psyche. He is not given the worst of the menial tasks. He is free, much as the creatures around him—that is, delicately watchful, not only of animals but of people, among whom life is not ranked subordination to authority. Conformity for him will be to social pressure and custom, not to force. All this is augured in the nonhuman world, not because he never sees dominant and subordinate animals, creatures killing others, or trees whose shade suppresses the growth of other plants, but because, reaching puberty, he is on the brink of a miracle of interpretation that will transform those things.

He will learn that his childhood experiences, though a comfort and joy, were a special language. Through myth and its ritual enactments, he is once again presented with that which he expects. Thenceforth natural things are not only themselves but a speaking. He will not put his delight in the sky and the earth behind him as a childish and irrelevant thing. The quests and tests that mark his passage in adolescent initiation are not intended to reveal to him that his love of the natural world was an illusion or that, having seemed only what it was, it in some way failed him. He will not graduate from that world but into its significance. So, with the end of childhood, he begins a lifelong study, a reciprocity with the natural world in which its depths are as endless as his own creative thought. He will not study it in order to transform its liveliness into mere objects that represent his ego, but as a poem, numinous and analogical, of human society.

The experience of such a world is initially that the mother is always there, following an easy birth in a quiet place, a presence in the tactile warmth of her body. For the infant there is a joyful comfort in being handled and fondled often, fed and cleaned as the body demands. From the start it is a world of variation on rhythms, the refreshment of hot and cold, wind like a breath in the face, the smell and feel of rain and snow, earth in hand and underfoot. The world is a soft sound-surround of gentle voices, human, cricket, and bird music. It is a pungent and inviting place with just enough bite that it says, "Come out, wake up, look, taste, and smell; now cuddle and sleep!"

There is a constancy of people, yet it is a world bathed in nonhuman forms, a myriad of figures, evoking an intense sense of their differences and similarities, the beckoning challenge of a lifetime. Speech is about that likeness and unlikeness, the coin of thought.

It is a world of travel and stop. At first the child fears being left and is bound by fear to the proximity of his mother and others. This interrupted movement sets the pace of his life, telling him gently that he is a traveler or visitor in the world. Its motion is like his own growth: as he gets older and as the cycle of group migrations is repeated, he sees places he has seen before, and those places seem less big and strange. The life of movement and rest is one of returning, and the places are the same and yet always more.

Now the child goes to the fringes of the camp to play at foraging. Play is an imitation, starting with simple fleeing and catching, going on to mimic joyfully the important animals, being them for a moment and then not being them, feeling as this one must feel and then that one, all tried on the self. The child sees the adults dancing the animal movements and does it too. Music itself has been there all the time, from his mother's song to the melodies of birds and the howls of wolves. The child already feels the mystery of kinship: likeness but difference.

The child goes out from camp with adults to forage and with playmates to imitate foraging. The adults show no anxiety in their hunting, only patience; one waits and watches and listens. Sometimes the best is not to be found, but there is always something. The world is all clues. There is no end to the subtlety and delicacy of the clues. The signs that reveal are always there. One has only to learn the art of reading them.

There is discomfort that cannot be avoided. The child sees with pride that he can endure it, that his body profits by it so that on beautiful days he feels wonderful. He witnesses sickness and death. But they are right as part of things and not really prevalent, for how could the little band of fifteen continue if there were dying every day?

The child is free. He is not asked to work. At first he can climb and splash and dig and explore the infinite riches about him. In time he increasingly wants to make things and to understand that which he cannot touch or change, to wonder about that which is unseen. His world is full of stories told, hearing of a recent hunt, tales of renowned events, and epics with layers of meaning. He has been bathed in voices of one kind or another always. Voices last only for their moment of sound, but they originate in life. The child learns that all life tells something and that all sound—from the frog calling to the sea surf—issues from a being kindred and significant to himself, telling some tale, giving some clue, mimicking some rhythm that hc should know. There is no end to what is to be learned.

At the end of childhood he comes to some of the most thrilling days of his life. The transition he faces will be experienced by body and ritual in concert. The childhood of journeying in a known world, scrutinizing and mimicking natural forms, and always listening has prepared him for a whole new octave in his being. The clock of his body permits it to be done, and the elders of his life will see that he is initiated. It is a commencement into a world foreshadowed by childhood: home, good, unimaginably rich, sometimes painful with reason, scrutable with care.

Western civilized cultures, by contrast, have largely abandoned the ceremonies of adolescent initiation that affirm the metaphoric, mysterious, and poetic quality of nature, reducing them to esthetics and amenities. But our human developmental program requires external models of order—if not a community of plants and animals, then words in a book, the ranks and professions of society, or the machine. If the ritual basis of the order-making metaphor is inadequate, the world can rigidify at the juvenile level of literalism, a boring place, which the adult will ignore as repetitive or exploit as mere substance.

In connecting ecological havoc to warpings of this ontogeny, I must also describe where I think such a healthy ontogeny should lead. I suspect I risk losing the whole case here: there are so many thoughtful portraits of maturity that I am bound to omit someone's criteria. My cues come from four men, none of whom can be blamed for my own knobby *bricolage*. I am captivated first by Erik Erikson's vision of a growing, clarifying sense of identity as the ultimate psychological goal of all the vicissitudes of infant, childhood, and adolescent passage making. Erikson sees the tasks of this sifting and differentiating as a meeting ground of heritable and experiential parts (together composing an epigenetic pattern) of the first two decades of life, not only leading to a heightened consciousness of personal uniqueness, but as the basis of the individuated personality itself. He sees ontogeny as a series of emergent tasks or critical-period concerns whose specific processes rise and fall in their intensity as surely as scenes in a drama, yet overlap like feathers on a bird. The successful traverse of Erikson's age-specific quest themes is essential.

To this idea of development as a crescendo of more detailed separateness Harold F. Searles has described in detail a vital counterplay: the concurrent growth of a conscious and unconscious web of connectedness analogous to the spread of kinship ties in human groups. Identity is not only a honing of personal singularity but a compounding wealth of ever more refined relationships between the person and increasingly differentiated parts of the rest of the world. Searles is emphatic that "world" means the nonhuman as well as the human, not only as the object of the linkages, but as the means and symbols of them.

The culmination of this refined difference-with-affinity is a firm ground of personal confidence and membership in its largest sense. And here I may rub the modern philosopher the wrong way, for I see in Carleton Coon's *The Hunting Peoples* the best description of that peculiar blend of loyalty and tolerance spanning the gap between separateness and belonging, forming a tapestry of ideas, feelings, speculation, experiences, and body that is Me, related to the multitude out there, from human brothers to the distant stars, that are the Other. Given this foundation and some intelligence, the years of tribal experience may produce wisdom, the most mature expression of which is the capacity to gently bring others forward to it, as far as they are capable.

Finally, that ethos of small-group tribal counsel with its advisory leadership, hospitality, and tolerance toward outsiders, perception of nature as a sacred language, and the quality of mentorship by which the young are coached was understood and exemplified by John Collier (d. 1968). His compassion for the American Indians was compounded by his knowledge of the caliber of those societies the whites had destroyed. In his eighty years he grew in sweetness and in tender understanding of his adversaries as well as his supporters both within and without tribal America. In age, he himself showed where the foundations of the first twenty years can lead. I think his maturity was most clearly manifest in his conviction of the goodness of creation and in his sense of being at home in the world.

The adult in full realization of his potential both uses and experiences the non-human world in characteristic ways, particularly in approaching it as both instrument and counterplayer, gift and home, and particularly not as an escape from or alternative to interpersonal and social relationships. To be fully mature, as Rollo May says, is to understand and to affirm limitations. There is also inherent in maturity an acceptance of ambiguity, of the tensions between the lust for omnipotence and the necessity to manipulate, between man as different and man as a kind of animal, and especially between a growing sense of the separateness of the self and kinship to the Other, achieved through an ever-deepening fullness of personal identity, defined by a web of relationship and metaphorical common ground. Harold Searles' remark is to the point: "It seems to me that the *highest* order of maturity is essential to the achievement of a reality relatedness with that which is *most unlike* oneself." Maturity emerges in midlife as the result of the demands of an innate calendar of growth phases, to which the human nurturers—parents, friends, teachers—have responded in season. It celebrates a central analogy of self and world in ever-widening spheres of meaning and participation, not an ever-growing domination over nature, escape in abstractions, or existential funk.

The human, twenty-year psychogenesis evolved because it was adaptive and beneficial to survival; its phases were specialized, integral to individual growth in the physical and cultural environments of the emergence of our species. And there is the rub—in those environments: small-group, leisured, foraging life-ways with natural surroundings. For us, now, that world no longer exists. The culmination of individual ontogenesis, characterized by graciousness, tolerance, and forbearance, tradition-bound to accommodate a mostly nonhuman world, and given to long, indulgent training of the young, may be inconsistent in some ways with the needs of society. In such societies—and I include ours—certain infantile qualities might work better: fear of separation, fantasies of omnipotence, oral preoccupation, tremors of helplessness, and bodily incompetence and dependence. Biological evolution is not involved, as it works much too slowly to make adjustments in our species in these ten millennia since the archaic foraging cultures began to be destroyed by their hostile, aggressive, better-organized, civilized neighbors. Programmed for the slow development toward a special kind of sagacity, we live in a world where that humility and tender sense of human limitation is no longer rewarded. Yet we suffer for the want of that vanished world, a deep grief we learn to misconstrue.

In the civilized world the roles of authority—family heads and others in power—were filled increasingly with individuals in a sense incomplete, who would in turn select and coach underlings flawed like themselves. Perhaps no one would be aware of such a trend, which would advance by pragmatic success across the generations as society put its fingers gropingly on the right moments in child nurturing by taking mothers off to work, spreading their attention and energy too thin with a houseful of babies, altering games and stories, manipulating anxiety in the child in a hundred possible ways. The transitory and normally healthful features of adolescence—narcissism, oedipal fears and loyalties, ambivalence and inconstancy, playing with words, the gang connection—might in time be honored in patriotic idiom and philosophical axiom. The primary impulses of infancy can be made to seem essential to belief, to moral superiority, masked by the psychological defenses of repression and projection. Over the centuries major institutions and metaphysics might finally celebrate attitudes and ideas originating in the normal context of immaturity, the speculative throes of adolescence, the Freudian psychosexual phases, or in even earlier neonatal or prenatal states.

Probably such ontogenetic crippling carries with it into adult life some traits that no society wants, but gets because they are coupled in some way with the childish will to destroy and other useful regressions, fellow travelers with ugly effects. Perhaps there is no way to perpetuate a suckling's symbiosis with mother as a religious ideal without dragging up painful unconscious memories of an inadequate body boundary or squeamishness about being cut loose.

In our time, youthfulness is a trite ideal, but even in those peasant and village societies where youth is "kept in its place," the traces of immaturity in cultural themes invite our scrutiny. That European medieval children were said to have no childhood as we understand it is suggestive of the omissions and constrictions that serve this socially adaptive crippling.

The person himself is, of course, caught between his inner calendar and the surgeries of society. His momentum for further growth may be twisted or amputated according to the hostilities, fears, or fantasies required of him, as his retardation is silently engineered to domesticate his integrity or to allow him to share in the collective dream of mastery.

I have chosen, arbitrarily, four historical periods in which to examine the mutilations of personal maturity as the vehicle of cultural progress and environmental decimation. They seem to offer especially good possibilities for seeking the impact of this process on the relation of man to nature. These are the earliest agriculture, the era of the desert fathers, the Reformation, and present-day industrial society. They are merely focal points in what may have been a continuing de-development, but I submit that they first began with a slight twist in the life of the child, with events that may only have marred his capacity for elderhood and judgment. If so, the history of Western man has been a progressive peeling back of the psyche, as if the earliest agriculture may have addressed itself to extenuation of adolescent concerns while the most modern era seeks to evoke in society at large some of the fixations of early natality—rationalized, symbolized, and disguised as need be. The individual growth curve, as described by Bruno Bettelheim, Jean Piaget, Erik Erikson, and others, is a biological heritage of the deep past. It is everyman's tree of life, now pruned by civic gardeners as the outer branches and twigs become incompatible with the landscaped order. The reader may extend that metaphor as he wishes, but I shall move to an animal image to suggest that the only society more frightful than one run by children, as in Golding's *Lord of the Flies*, might be one run by childish adults.

The Futility of Global Thinking

Wendell Berry

It is conventional at graduation exercises to congratulate the graduates. Though my good wishes for your future could not be more fervent, I think I will refrain from congratulations. This, after all, is your commencement, and a beginning is the wrong time for congratulations. What I want to attempt instead is to say something useful about the problems and the opportunities that lie ahead of your generation and mine.

Toward the end of *As You Like It*, Orlando says: "I can live no longer by thinking." He is ready to marry Rosalind. It is time for incarnation. Having thought too much, he is at one of the limits of human experience, or of human sanity. If his love does put on flesh, we know he must sooner or later arrive at the opposite limit, at which he will say, "I can live no longer without thinking." Thought—even consciousness—seems to live between these limits: the abstract and the particular, the word and the flesh.

Harper's, September 1989. (Adapted from "Word and Flesh," a commencement address given by Wendell Berry in June at the College of the Atlantic in Bar Harbor, Maine.)

All public movements of thought quickly produce a language that works as a code, useless to the extent that it is abstract. It is readily evident, for example, that you can't conduct a relationship with another person in terms of the rhetoric of the civil rights movement or the women's movement—as useful as those rhetorics may initially have been to personal relationships.

The same is true of the environment movement. The favorite adjective of this movement now seems to be *planetary*. This word is used, properly enough, to refer to the interdependence of places, and to the recognition, which is desirable and growing, that no place on the earth can be completely healthy until all places are.

But the word *planetary* also refers to an abstract anxiety or an abstract passion that is desperate and useless exactly to the extent that it is abstract. How, after all, can anybody—any particular body—do anything to heal a planet? Nobody can do anything to heal a planet. The suggestion that anybody could do so is preposterous. The heroes of abstraction keep galloping in on their white horses to save the planet—and they keep falling off in front of the grandstand.

What we need, obviously, is a more intelligent—which is to say, a more accurate—description of the problem. The description of a problem as planetary arouses a motivation for which, of necessity, there is no employment. The adjective *planetary* describes a problem in such a way that it cannot be solved. In fact, though we now have serious problems nearly everywhere on the planet, we have no problem that can accurately be described as planetary. And, short of the total annihilation of the human race, there is no planetary solution.

There are also no national, state, or county problems, and no national, state, or county solutions. That will-o'-the-wisp, the large-scale solution to the large-scale problem, which is so dear to governments, universities, and corporations, serves mostly to distract people from the small, private problems that they may, in fact, have the power to solve.

The problems, if we describe them accurately, are all private and small. Or they are so initially.

The problems are our lives. In the "developed" countries, at least, the large problems occur because all of us are living either partly wrong or almost entirely wrong. It was not just the greed of corporate shareholders and the hubris of corporate executives that put the fate of Prince William Sound into one ship; it was also our demand that energy be cheap and plentiful.

The economies of our communities and households are wrong. The answers to the human problems of ecology are to be found in economy. And the answers to the problems of economy are to be found in culture and in character. To fail to see this is to go on dividing the world falsely between guilty producers and innocent consumers.

The planetary versions—the heroic versions—of our problems have attracted great intelligence. Our problems, as they are caused and suffered in our lives, our households, and our communities, have attracted very little intelligence.

There are some notable exceptions. A few people have learned to do a few things better. But it is discouraging to reflect that, though we have been talking about most of our problems for decades, we are still mainly *talking* about them. The civil rights movement has not given us better communities. The women's movement has not given us better marriages or better households. The environment movement has not changed our parasitic relationship to nature.

We have failed to produce new examples of good home and community economics, and we have nearly completed the destruction of the examples we once had. Without examples, we are left with theory and the bureaucracy and the meddling that come with theory. We change our principles, our thoughts, and our words, but these are changes made in the air. Our lives go on unchanged.

For the most part, the subcultures, the countercultures, the dissenters, and the opponents continue mindlessly—or perhaps just helplessly—to follow the pattern of the dominant society in its extravagance, its wastefulness, its dependencies, and its addictions. The old problem remains: How do you get intelligence *out* of an institution or an organization?

My small community in Kentucky has lived and dwindled for a century at least under the influence of four kinds of organizations: governments, corporations, schools, and churches—all of which are distant (either actually or in interest), centralized, and consequently abstract in their concerns.

Governments and corporations (except for employees) have no presence in our community at all, which is perhaps fortunate for us, but we nevertheless feel the indifference or the contempt of governments and corporations for communities such as ours.

We have had no school of our own for nearly thirty years. The school system takes our young people, prepares them for "the world of tomorrow," which it does not expect to take place in any rural area, and gives back expert (that is, extremely generalized) ideas.

The church is present in the town. We have two churches. But both have been used by their denominations, for almost a century, to provide training and income for student ministers, who do not stay long enough even to become disillusioned.

For a long time, then, the minds that have most influenced our town have not been *of* the town and so have not tried even to perceive, much less to honor, the good possibilities that are there. They have not wondered on what terms a good and conserving life might be lived there. In this, my community is not unique but is like almost every other neighborhood in our country and in the "developed" world.

The question that must be addressed, therefore, is not how to care for the planet but how to care for each of the planet's millions of human and natural neighborhoods, each of its millions of small pieces and parcels of land, each one of which is in some precious way different from all the others. Our understandable wish to preserve the planet must somehow be reduced to the scale of our competence—that is, to the wish to preserve all of its humble households and neighborhoods.

What can accomplish this reduction? I will say again, without overweening hope but with certainty nonetheless, that only love can do it. Only love can bring intelligence out of the institutions and organizations, where it aggrandizes itself, into the presence of the work that must be done.

Love is never abstract. It does not adhere to the universe or the planet or the nation or the institution or the profession but to the singular sparrows of the street, the lilies of the field, "the least of these my brethren." Love is not, by its own desire, heroic. It is heroic only when compelled to be. It exists by its willingness to be anonymous, humble, and unrewarded.

The older love becomes, the more clearly it understands its involvement in partiality, imperfection, suffering, and mortality. Even so, it longs for incarnation. It can live no longer by thinking.

And yet to put on flesh and do the flesh's work, it must think.

In his essay on Kipling, George Orwell wrote:

> All left-wing parties in the highly industrialized countries are at bottom a sham, because they make it their business to fight against something which they do not really wish to destroy. They have internationalist aims, and at the same time they struggle to keep up a standard of life with which those aims are incompatible. We all live by robbing Asiatic coolies, and those of us who are "enlightened" all maintain that those coolies ought to be set free; but our standard of living, and hence our "enlightenment," demands that the robbery shall continue.

This statement of Orwell's is clearly applicable to our situation now; all we need to do is change a few nouns. The religion and the environmentalism of the highly industrialized countries are at bottom a sham, because they make it their business to fight against something that they do not really wish to destroy. We all live by robbing nature, but our standard of living demands that the robbery shall continue.

We must achieve the character and acquire the skills to live much poorer than we do. We must waste less. We must do more for ourselves and each other. It is either that or continue merely to think and talk about changes that we are inviting catastrophe to make.

The great obstacle is simply this: the conviction that we cannot change because we are dependent upon what is wrong. But that is the addict's excuse, and we know that it will not do.

How dependent, in fact, are we? How dependent are our neighborhoods and communities? How might our dependences be reduced? To answer these questions will require better thoughts and better deeds than we have been capable of so far.

We must have the sense and the courage, for example, to see that the ability to transport food for hundreds or thousands of miles does not necessarily mean that we are well off. It means that the food supply is more vulnerable and more costly than a local food supply would be. It means that consumers do not control or influence the healthfulness of their food supply and that they are at the mercy of the people who have the control and influence. It means that, in eating, people are using large quantities of petroleum that other people in another time are almost certain to need.

I am trying not to mislead you, or myself, about our situation. I think that we have hardly begun to realize the gravity of the mess we are in.

Our most serious problem, perhaps, is that we have become a nation of fantasists. We believe, apparently, in the infinite availability of finite resources. We persist in land-use methods that reduce the potentially infinite power of soil fertility to a finite quantity, which we then proceed to waste as if it were an infinite quantity. We have an economy that depends not upon the quality and quantity of necessary goods and services but on the behavior of a few stockbrokers. We believe that democratic freedom can be preserved by people ignorant of the history of democracy and indifferent to the responsibilities of freedom.

Our leaders have been for many years as oblivious to the realities and dangers of their time as were George III and Lord North. They believe that the difference between war and peace is still the overriding political difference—when, in fact, the difference has diminished to the point of insignificance. How would you describe the difference between modern war and modern industry—between, say, bombing and strip mining, or between chemical warfare and chemical manufacturing? The difference seems to be only that in war the victimization of humans is directly intentional and in industry it is "accepted" as a "trade-off."

Were the catastrophes of Love Canal, Bhopal, Chernobyl, and the *Exxon Valdez* episodes of war or of peace? They were, in fact, peacetime acts of aggression, intentional to the extent that the risks were known and ignored.

We are involved unremittingly in a war not against "foreign enemies" but against the world, against our freedom, and indeed against our existence. Our so-called industrial accidents should be looked upon as revenges of Nature. We forget that Nature is necessarily party to all our enterprises and that she imposes conditions of her own.

Now she is plainly saying to us, "If you put the fates of whole communities or cities or regions or ecosystems at risk in single ships or factories or power plants, then I will furnish the drunk or the fool or the imbecile who will make the necessary small mistake."

And so, graduates, my advice to you is simply my hope for us all:

Beware the justice of Nature.

Understand that there can be no successful human economy apart from Nature or in defiance of Nature.

Understand that no amount of education can overcome the innate limits of human intelligence and responsibility. We are not smart enough or conscious enough or alert enough to work responsibly on a gigantic scale.

In making things always bigger and more centralized, we make them both more vulnerable in themselves and more dangerous to everything else. Learn, therefore, to prefer small-scale elegance and generosity to large-scale greed, crudity, and glamour.

Make a home. Help to make a community. Be loyal to what you have made.

Put the interest of the community first.

Love your neighbors—not the neighbors you pick out, but the ones you have.

Love this miraculous world that we did not make, that is a gift to us.

As far as you are able make your lives dependent upon your local place, neighborhood, and household—which thrive by care and generosity—and independent of the industrial economy, which thrives by damage.

Find work, if you can, that does no damage. Enjoy your work. Work well.

From Global Bioethics: Building on the Leopold Legacy

Van Rensselaer Potter

The Cancer Analogy

On December 28, 1954, the American Association for the Advancement of Science held a symposium on "Population Problems," at which Dr. Alan Gregg, vice-president of the Rockefeller Foundation (1951–56), came up with a startling idea: the thought that the human species is to the planet Earth what a cancer is to an individual human being. As a cancer specialist I was aware of the many contributing lines of thought and so was not altogether surprised to note the same idea proposed by another eminent biologist, Professor Norman J. Berrill of McGill University, in his superb book

Global Bioethics: Building on the Leopold Legacy (East Lansing: Michigan State University Press, 1988).

Man's Emerging Mind.[1] It was published in 1955, the same year that Gregg's symposium paper appeared in *Science.*[2] The remarks by these two men of science suggest that the effect of an ever-expanding human population on the carrying-capacity of the planet Earth bears examination. We do would well to examine their words.

EARTH AS ORGANISM

Gregg espoused an idea that was clearly enunciated in 1949 by Aldo Leopold, who referred to "land the collective organism" and stated, "Land, then, is not merely soil; it is a fountain of energy flowing through a circuit of soils, plants, and animals. Food chains are the living channels which conduct energy upward; death and decay return it to the soil."[3] Leopold also anticipated Gregg and Berrill when he remarked, "This almost worldwide display of disorganization in the land seems to be similar to disease in an animal except that it never culminates in complete disorganization or death. The land recovers, but at some reduced level of complexity, and with a reduced carrying capacity for people, plants, and animals." (*Almanac*, 297).

Alan Gregg proposed similar ideas in his aforementioned symposium paper: "If we regard the different forms of plant and animal life in the world as being so closely related to and dependent on one another that they resemble different types of cells in a total organism, then we may, for the sake of a hypothesis, consider the living world as an organism." He went on to say:

> What would we think if it became evident that within a very brief period in the history of the world some one type of its forms of life had increased greatly at the expense of other types of life? In short, I suggest, as a way of looking at the population problem, that there are some interesting analogies between the growth of the human population of the world and the increase of cells observable in neoplasms. To say that the world has cancer, and that the cancer cell is man has neither experimental proof nor the validation of predictive accuracy; but I see no reason that instantly forbids such a speculation. . . . Cancerous growths demand food; but, so far as I know, they have never been cured by getting it. . . . How nearly the slums of our great cities resemble the necrosis of tumors raises the whimsical query: which is the more offensive to decency and beauty,

slums or the fetid detritus of a growing tumor? . . . If Copernicus helped astronomy by challenging the geocentric interpretation of the universe, might it not help biology to challenge the anthropocentric interpretation of nature?

Norman Berrill was in complete agreement with Leopold and Gregg. In "The Human Crop," chapter 17 of *Man's Emerging Mind*, he discussed the issue at length:

> Directly or indirectly there has been a monumental and increasingly extensive conversion of the planet's living potential from the diverse many to the all-consuming one. In terms of our comparison, the virgin prairie with its stable mixture of grasses and flowers has become almost entirely corn, with a few weeds and some blowing dust. All that can be transformed into human protoplasm is being transformed, and anything that stands in the way is pushed against the wall. . . . So far as the rest of nature is concerned we are like a cancer whose strange cells multiply without restraint, ruthlessly demanding the nourishment that all of the body has need of. The analogy is not far-fetched for cancer cells no more than whole organisms know when to stop multiplying, and sooner or later the body or the community is starved of support and dies. (209–10)

Berrill saw three possible responses to the problem of overpopulation:

> One is that we can increase our resources indefinitely to keep pace with the increasing population, which I have tried to show is impossible. Another is that we employ our collective intelligence and keep our numbers within reasonable bounds, while the third is the pessimistic one that human beings are not intelligent enough as a whole to control their own fertility and will always press hard against the ragged fringe of sustenance . . . that always the more fertile or the more prolific human strains or races will outbreed the rest, that population control by any group sooner or later seals its own doom, with those who retain an uncontrollable breeding instinct taking its place. (220–21)

The analogy that sees Earth as an organism with all living species as cells in that organism is not complete in detail because the various species exist by and large by consuming the bodies, living or dead, of other species. In contrast, the cells that form an organism in the human body do not live by consuming other cells in the community, although the total human organism does depend on the intake of plant and animal species as food. Nevertheless, the proliferation of cell types within a human organism is exquisitely regulated by feedback mechanisms of great complexity, and the same can be said for the proliferation of living species on the planet Earth. In either case, when "some one type of its forms of life had increased greatly at the expense of other types of life," to use Gregg's words, we must conclude that the natural feedback mechanisms, evolved over millions of years, have broken down. It becomes clear that in either case the result is brought about by a great excess of births over the number of deaths in a given time interval. If the human species is to survive and prosper, it is essential that we must control not only nuclear armaments but also human fertility and the tendency to crowd out or destroy other forms of life. This statement of "what we must do" is merely an extension of the concluding Leopold Paradigm: "A thing [referring to decent land use] is right when it tends to preserve the integrity, stability, and beauty of the biotic community. It is wrong when it tends to do otherwise."

THE ISSUE OF SURVIVAL

It was perhaps Garrett Hardin more than any other who independently developed the equivalent of the Leopold paradigms and realized that this led directly to the issue of fertility control for the human population. In 1968 he wrote "The Tragedy of the Commons" in which he concluded that "freedom to breed is intolerable."[4] In 1972 he went beyond Leopold when he wrote, "With the flowering of concern for environmental quality and the growth of theory in ecology the time is now ripe, I think, for a concerted attack on the population-environment-quality complex. I think it is almost time to grasp the nettle of population control, which we sometime must, if we are to survive with dignity."[5] With that statement he began what I propose to continue, that is, the adoption of the criterion of survival as a guide for action, and the discussion of what kind of survival we should advocate.

Like Garrett Hardin, Eugene P. Odum, co-author of *Fundamentals of Ecology,* was concerned with the relation between population and survival. Quoting extensively from Leopold's views on the need to extend ethics to the relation of man to the natural environment, Odum wrote: "We can also present strong scientific and technological reasons for the proposition that such a major extension of the general theory of ethics is now necessary for human survival."[6] Like Gregg and Berrill he noted the cancer analogy: "Growth beyond the optimum becomes cancer. Cancer is an ever-present threat to any mature system and must constantly be guarded against."[7] In proposing "The Emergence of Ecology as a New Integrative Discipline" in 1977,[8] he quoted Alex Novikoff on reductionism and holism: "Equally essential for the purposes of scientific analysis are both the isolation of parts of a whole and their integration into the structure of the whole. . . . The consideration of one to the exclusion of the other acts to retard the development of biological and sociological sciences."[9] Odum concluded: "To achieve a truly holistic or ecosystematic approach, not only ecology but other disciplines in the natural, social, and political sciences as well must emerge to new hitherto unrecognized and unresearched levels of thinking and action" ("Emergence of Ecology," 1291), echoing Hardin's plea for "a concerted attack on the population-environment-quality complex."

When Richard Falk of Princeton University wrote *This Endangered Planet: Prospects and Proposals for Human Survival* (1971), he discussed the four dimensions of planetary danger as: (1) the war system; (2) population pressure; (3) insufficiency of resources; and (4) environmental overload. Following Paul Shepard, who in turn had paid tribute to Rachel Carson and Aldo Leopold, Falk reflected Odum's views when he wrote:

> Such a posture of concern and position makes of human ecology a kind of ethics of survival. It is a science that relies on careful procedures of inquiry, data collection, and detailed observation as the basis of inference, explanation, and prediction. But it also involves a moral commitment to survival and to the enhancement of the natural habitat of man.[10]

All of the above authors have seen a link between ecology, population pressure, and human survival. They have in general not considered what kind of survival, although Hardin spoke of surviving "with dignity" and Berrill visualized humankind always pressing hard against "the ragged fringe of sustenance." Unlike the vast majority of the human species, they are aware that humankind has no guarantee of survival, that survival cannot be assumed.

References

1. Norman J. Berrill, *Man's Emerging Mind* (New York: Dodd, Mead and Co., 1955).
2. Alan Gregg, "A Medical Aspect of the Population Problem," *Science* 121 (1955): 681–82.
3. Aldo Leopold, *Sand County Almanac*, 1987 ed., 216. The page references herein are from the 1987 edition (see chapter 1, n. 1).
4. Garrett Hardin, "The Tragedy of the Commons," *Science* 162 (1968): 1243–48. This essay has been reprinted in countless subsequent publications.
5. Garrett Hardin, *Exploring New Ethics for Survival* (New York: The Viking Press, 1972; reprint ed., Baltimore: Penguin Books, 1973).
6. Eugene P. Odum and H. T. Odum, *Fundamentals of Ecology*, 3d ed. (Philadelphia: Saunders, 1971), 10.
7. Eugene P. Odum, "Environmental Ethic and Attitude Revolution," in *Philosophy and Environmental Crisis*, ed. Wm. T. Blackstone (Athens: University of Georgia Press, 1974), 14. In this article Odum quoted from Leopold's "Land Ethic."
8. Eugene P. Odum, "The Emergence of Ecology as a New Integrative Discipline," *Science* 195 (1977): 1289–93.
9. Alex B. Novikoff, "The Concept of Integrative Levels and Biology," *Science* 101 (1945): 209–15.
10. Richard Falk, *This Endangered Planet: Prospects and Proposals for Human Survival* (New York: Vintage Books, Random House, 1971), 187.

Why We Have Failed

Barry Commoner

In 1970, in response to growing concern, the U.S. Congress began a massive effort to undo the pollution damage of the preceding decades. In short order, legislators in Washington passed the National Environmental Protection Act (NEPA) and created the Environmental Protection Agency (EPA) to administer it. These two events are the cornerstone of what is indisputably the world's most vigorous pollution control effort, a model for other nations and a template for dozens of laws and amendments passed since. Now, nearly twenty years later, it is time to ask an important and perhaps embarrassing question: how far have we progressed toward the goal of restoring the quality of the environment?

The answer is indeed humbling. Apart from a few notable exceptions, environmental quality has improved only slightly, and in some cases worsened. Since 1975, emissions of sulfur dioxide and carbon monoxide are down by about 19 percent, but nitrogen oxides are up about 4 percent. Overall improvement in major pollutants amounts to only about 15 to 20 percent, and the rate of improvement has actually slowed considerably.

Greenpeace, September/October 1989.

There are several notable and heartening exceptions. Pollution levels of a few chemicals—DDT and PCBs in wildlife and people, mercury in the fish of the Great Lakes, strontium 90 in the food chain and phosphate pollution in some local rivers—have been reduced by 70 percent or more. Levels of airborne lead have declined more than 90 percent since 1975.

The successes explain what works and what does not. Every success on the very short list of significant environmental quality improvements reflects the same remedial action: production of the pollutant has been stopped. DDT and PCB levels have dropped because their production and use have been banned. Mercury is much less prevalent because it is no longer used to manufacture chlorine. Lead has been taken out of gasoline. And strontium has decayed to low levels because the United States and the Soviet Union had the good sense to stop the atmospheric nuclear bomb tests that produced it.

The lesson is plain: pollution prevention works; pollution control does not. Only where production technology has been changed to eliminate the pollutant has the environment been substantially improved. Where it remains unchanged, where an attempt is made to trap the pollutant in an appended control device—the automobile's catalytic converter or the power plant's scrubber—environmental improvement is modest or nil. When a pollutant is attacked at the point of origin, it can be eliminated. But once it is produced, it is too late.

PROGRESS AND POLLUTION

Most of our environmental problems are the inevitable result of the sweeping technological changes that transformed the U.S. economic system after World War II: the large, high-powered cars; the shift from fuel-efficient railroads to gas-guzzling trucks and cars; the substitution of fertilizers for manure and crop rotation and of toxic synthetic pesticides for ladybugs and birds.

By 1970, it was clear that these technological changes were the root cause of environmental pollution. But the environmental laws now in place do not address the technological origin of pollutants. I remember the incredulity in Senator Edmund Muskie's voice during NEPA hearings when he asked me whether I was really testifying that the technology that generated postwar economic progress was also the cause of pollution. I was.

Because environmental legislation ignored the origin of the assault on environmental quality, it has dealt only with its subsequent problems—in effect defining the disease as a collection of symptoms. As a result, all environmental legislation mandates only palliative measures. The notion of preventing pollution—the only measure that really works—has yet to be given any administrative force.

The goal established by the Clean Air Act in 1970 could have been met if the EPA had confronted the auto industry with a demand for fundamental changes in engine design, changes that were practical and possible. And had American farmers been required to reduce the high rate of nitrogen fertilization, nitrate water pollution would now be falling instead of increasing.

If the railroads and mass transit were expanded, if the electric power system were decentralized and increasingly based on cogenerators and solar sources, if American homes were weatherized, fuel consumption and air pollution would be sharply reduced. If brewers were forbidden to put plastic nooses on six-packs of beer, if supermarkets were not allowed to wrap polyvinyl chloride film around everything in sight, if McDonald's restaurants could rediscover the paper plate, if the use of plastics was cut back to those things considered worth the social costs (say, artificial hearts or video tape), then we could push back the petrochemical industry's toxic invasion of the biosphere.

Of course, all this is easier said than done. I am fully aware that what I am proposing is no small thing. It means that sweeping changes in the major systems of production—agriculture, industry, power production, and transportation—would be undertaken for a social purpose: environmental improvement. This represents social (as contrasted with private) governance of the means of production—an idea that is so foreign to what passes for our national ideology that even to mention it violates a deep-seated taboo.

The major consequence of this powerful taboo is the failure to reach the goals in environmental quality that motivated the legislation of the 1970s.

RISK AND PUBLIC MORALITY

In the absence of a prevention policy, the EPA adopted a convoluted pollution control process. First, the EPA must estimate the degree of harm represented by different levels of the numerous environmental pollutants. Next, some "acceptable" level of harm is chosen (for example, a cancer risk of one in a million) and emission and/or ambient concentration standards that can presumably achieve that risk level are established.

Polluters are then expected to respond by introducing control measures (such as automobile exhaust catalysts or power plant stack scrubbers) that will bring emissions to the required levels. If the regulation survives the inevitable challenges from industry (and in recent years from the administration itself), the polluters will invest in the appropriate control systems. Catalysts are attached to cars, and scrubbers to the power plants and trash-burning incinerators. If all goes well—and it frequently does not—at least some areas of the country and some production facilities are then in compliance with the regulation.

The net result is that an "acceptable" pollution level is frozen in place. Industry, having invested heavily in equipment designed to reach just the required level, is unlikely to invest in further improvements.

Clearly, this process is the opposite of the preventive approach to public health. It strives not for the continuous improvement of environmental health, but for the social acceptance of some, hopefully low, risk to health. By contrast, the preventive approach aims at progressively reducing the risk to health. It does not mandate some socially convenient stopping point. The medical professions, after all, did not decide that the smallpox prevention program could stop when the risk reached one in a million. They kept on, and the disease has now been wiped out worldwide.

How do you decide when to stop, where to set the standard for acceptable pollution? The current fashion is to submit the question to a risk/benefit analysis. Since the pollutants' ultimate effect can often be assessed by the number of lives lost, the risk/benefit analysis requires that a value be placed on human life. Such reckoning often bases that value on lifelong earning power, so that a poor person's life is worth less than a rich person's. So, on the risk/benefit scale, the poor can be exposed to more pollution than the rich. In fact, this is what is happening in the United States: the burden of an environmental risk—say the siting of a municipal incinerator or a hazardous waste landfill—falls disproportionally on poor people, who lack the political and financial clout to deter the risk.

In this way, risk/benefit analysis—a seemingly straightforward numerical computation—conceals a profound, unresolved moral question: should poor people be subjected to a more severe environmental burden than rich people, simply because they lack the resources to evade it? Since in practice the risk/benefit equation masquerades as science, it relieves society of the duty to confront this question. One result of failing to adopt the preventive approach is that the regulatory agencies have been driven into positions that seriously diminish the force of social morality.

THE REAL SOLUTION

The fate of Alar, the pesticide used to enhance the marketability of apples, provides a recent instructive example of what prevention means. Like many other petrochemical products, Alar poses a health risk. It has been proven to induce cancer in test animals. As in many other such cases, a debate has flourished over the extent of the hazard to people, especially children, and over what standards should be applied to limit exposure to "acceptable" levels.

In June, Alar broke out of the pattern when the manufacturer, Uniroyal, decided that regardless of the toxicological uncertainties, Alar would be taken off the market. They acted simply because parents were unhappy about raising their children on apple juice that represented *any* threat to their health. Food after all, is supposed to be good for you.

This is a clear-cut example of the benefits of prevention, as opposed to control. Pollution prevention means identifying the source of the pollutant in the production process, eliminating it from that process and substituting a more environmentally benign method of production. This differentiates it from source reduction (reducing the amount of the pollutant produced, either through altering processes or simple housekeeping) and pollution control. Once a pollutant is eliminated, the elaborate system of risk assessment, standard setting, and the inevitable debates and litigation become irrelevant.

Instituting the practice of prevention rather than control will require the courage to challenge the taboo against questioning the dominance of private interests over the public interest. But I suggest that we begin with an open public discussion of what has gone wrong, and why. That is the necessary first step on the road toward realizing the nation's unswerving goal—restoring the quality of the environment.

From One Life at a Time, Please

Edward Abbey

Arizona: How Big Is Big Enough?

Governor Bruce Babbit tells us that by the year 2000, only sixteen years from now, Arizona will gain 2 million new residents, that Phoenix will become another Houston and Tucson another Phoenix, and that we will have an additional 1 million automobiles crowding our streets and highways. Tucson Mayor Lew Murphy—unable to conceal his smirking glee—predicts that Tucson will become, within twenty years, a 450-square-mile urbanized area. Most of our reigning bankers, economists, and developers keep shelling us with a similar barrage of thundering numbers. This, our leaders tell us, is good news. Growth is good, they say, reciting

One Life at a Time, Please (New York: Henry Holt, 1988).

like an incantation the prime article of faith of the official American religion: Bigger is better and best is biggest. Growth, they tell us, means more jobs, more bank accounts, more cars, more people, leading in turn to the demand for more jobs, more economic expansion, more industrial development. Where, when, and how is this spiraling process supposed to reach a rational end—a state of stability, sanity, and equilibrium?

When and if Arizona becomes like Southern California or South Central Texas or the Baltimore-to-Boston megalopolis, will people like Murphy and Babbitt and the corporation executives from whom they take their instructions then be satisfied? Or will our children be faced, once again, with new and greater demands for still more growth?

Already looking beyond the completion of the Central Arizona Project, Babbitt speaks of mining the ground water of Western Arizona. For what purpose? Why, to provide the essential liquid element for further growth. And when that supply is played out, as it is already playing out in Southern Arizona and the high plains of Texas, then what? The answer is easy to foresee: A great clamor from our Southwestern politicians to desalinate the Sea of Cortez, to import icebergs from Antarctica, to divert first the Columbia and then the Yukon rivers into the drainages of the Colorado.

Viewing it in this way, we can see that the religion of endless growth—like any religion based on blind faith rather than reason—is a kind of mania, a form of lunacy, indeed a disease. And the one disease to which the growth mania bears an exact analogical resemblance is cancer. Growth for the sake of growth is the ideology of the cancer cell. Cancer has no purpose but growth; but it does have another result—the death of the host.

But all this is mere futurology, like astrology and computerology and technology, only one of the many commercialized superstitions of our time. We need not look years ahead but simply look at the present to weigh the comparative advantages and disadvantages of industrial growth. We need only consider Phoenix and Tucson and decide which of the two is the more attractive, which is the better place to live in, make a living in, raise a family in.

It should be clear to everyone by now that crude numerical

growth does not solve our chronic problems of unemployment, welfare, crime, traffic, filth, noise, squalor, the pollution of our air, the poisoning of our water, the corruption of our politics, the debasement of the school system (hardly worthy of the name "education"), and the general loss of popular control over the political process—where money, not people, is now the determining factor.

Far from solving such problems, industrial expansion and population growth only make them worse. Does Houston really provide us with a model to aspire toward? Or Chicago? New York? Los Angeles? Miami? Or maybe Mexico City?

Ah well, say our alleged leaders, in response to this sort of argument, we face a challenge. Our politicians love that word. And the challenge, they tell us, is to accommodate ourselves to endless growth without sacrificing the quality of the Arizona environment, without losing the bright skies, the bracing air, the open space, the abundant wildlife, the desert plant life, the sheer delight of physical freedom, all of those good and unique and irreplaceable things that are in sum what attracted most of us to Arizona in the first place.

We can have it both ways, they say. We can enjoy our cake and at the same time destroy it, grind it to bits in the urbanizing, industrializing mill, and transform what we prize into boom time, if temporary, jobs for thousands and fat bank accounts for the tiny but powerful minority of land speculators, tract-slum builders, bankers, car dealers, and shopping-mall hustlers who stand to profit from what they call growth.

This argument hardly requires an answer. The so-called challenge is a plain lie. All industrial development involves a trade-off: in order to make room for more growth, we must give up the very qualities that make a high standard of civilized human life still possible in Arizona—as contrasted to, say, the frantic, crowded, substandard life of California's Silicon Valley. (Do you really want to live in a place where the microchip is the highest object of human desire?)

And now we come to the final argument of growth zealots. Growth, they say, is inevitable. There is nothing, they say, that we can do about it. There is no constitutional means by which we can prevent 2 million more flatlanders from invading Arizona in the next sixteen years.

This is the baldest lie of them all. Nothing is inevitable but death, taxes, and the insolent dishonesty (as Mark Twain said) of elected officials.

The fact is that the fungoid growth of Arizona in recent years has been the result of deliberate policy. The only purpose of the CAP is to make possible the continued growth of Central and Southern Arizona. The same is true of the Palo Verde nuclear power plant and the various new coal-burning plants now polluting the public air.

The only purpose of the state's pro-business, anti-labor position is to lure industry here—and if this policy causes misery and hardship elsewhere, that is of no concern to our leaders. The only purpose of freeway projects, highway building, river-damming, pro-development rezoning, and opposition to wilderness preservation, naming but a few measures, is to make possible, encourage, and create the runaway growth that enriches a few and gradually impoverishes the rest of us.

If we in Arizona did our part in keeping American industry where it now is, we would also help keep Arizona what it is. People follow industry, high-technology or otherwise, not because they enjoy being uprooted from their homes but out of painful economic need.

Does my attitude seem selfish? Of course it is. I have lived in the Southwest since 1947—forty years—most of my life. During most of those years, I survived on part-time work and the precarious existence of a free-lance writer, usually on an income below the official U.S. government poverty line. I did it because I love Arizona and the Southwest, preferably as it was but even as it is. And because I love it, I do not want to see our state become one more high-tech slum like California or a wasteland of space-age sleaze like Texas. I have a sneaking suspicion there's about a million other Arizonans who feel the same way I do.

We cannot creep from quantity to quality. It's high time we told the little cabal who run this state that Tucson is big enough, Phoenix is big enough, Arizona is big enough. What we need is not more growth but more democracy—and democracy, some other old-timers may recall, means government by the people. *By* the people.

Every River I Touch Turns to Heartbreak

Edward Abbey

Every precious moment entails every other. Each sacred place suggests the immanent presence of all places. Each man, each woman, exemplifies all humans. The bright faces of my companions, here, now, on this Rio Dolores, this River of Sorrows, somewhere in the melodramatic landscape of southwest Colorado, break my heart—for in their faces, eyes, vivid bodies in action, I see the hope and joy and tragedy of humanity everywhere.

Here we go again, down one more condemned river. Our foolish rubber rafts nose into the channel and bob on the current. Brown waves glitter in the sunlight. The long oars of the boatpeople—young women, young men—bite into the heavy water. Snow melt from the San Juan Mountains creates a river in flood, and the cold waters slide past the willows, hiss upon the gravel bars, thunder and roar among the rocks in a foaming chaos of exaltation.

Call me Jonah. I should have been a condor sailing high above the gray deserts of the Atacama. I should have stayed in Hoboken when I had the chance. Every river I touch turns to heartbreak.

Environmental Action, "Visions," May/June 1985.

Floating down a portion of Rio Colorado in Utah on a rare month in spring, twenty-two years ago, a friend and I found ourselves passing through a world so beautiful it seemed and had to be . . . eternal.

Such perfection of being, we thought, these glens of sandstone could not possibly be changed. The philosopher and the theologians have agreed that the perfect is immutable—that which cannot alter and cannot ever be altered.

They were wrong. We were wrong. Glen Canyon was destroyed. Everything changes, and nothing is more vulnerable than the beautiful.

Why yes, the Dolores is scheduled for damnation. Only a little dam, say the politicans, one little earth-fill dam to irrigate the sorghum and alfalfa plantations, and then, most likely, to supply the industrial parks and syn-fuel factories of Cortez, Shithead Capital of Dipstick, Colorado. True, only a little dam. But dammit, it's only a little river.

Forget it. Write it off. Fix your mind on the feel of the oars in your hands. Watch out for that waterlogged fir tree. Follow that young lady boatman ahead, she's been down this one before. . . .

A good boatman must know when to act, when to react, and when to rest. I lean on the oars, lifting them like bony wings from the water and ignore the whining and mewling from two passengers behind me. . . .

I think of lunch: Tuna from a tin, beslobbered with mayonnaise. Fig Newtons and Oreo cookies. A thick-skin Sunkist orange . . . Our world is so full of beautiful things: Fruit and ideas and woman and banjo music and onions with purple skins. A virtual Paradise. But even Paradise can be damned, flooded, mucked up by fools in pursuit of paper profits and plastic happiness.

My thoughts wander to Mark Dubois. Talk about the *right stuff.* That young man chained himself to a rock, in a hidden place known only to a single friend, in order to save a river he had learned to know and love too much: The Stanislaus in northern California. Mark Dubois put his life on the rock, below high-water line, and drove half the officialdom of California and the Army Corps of Engineers into exasperated response, forcing them to halt the filling of what they call the New Melones Dam.

In comparing U.S. government functionaries to those of the Soviet Union or China or Brazil or Chile, we are obliged to give our own a degree of credit: They are still reluctant to sacrifice human lives to industrial purposes in the full glare of publicity. (Why we need a free press.) But I prefer to give my thanks direct to people like Mark Dubois, whose courage, in serving a cause worthy of service, seems to me of much more value than that of our astronauts and cosmonauts and other assorted technetronic whatnots: Dropouts, all of them, from the real world of earth, rivers, life. . . .

Be of good cheer. All may yet be well. There's many a fork, I think, on the road from here to destruction. Despite the jet-set androids who visit our mountain West on their cyclic tours from St. Tropez to Key West to Vail to Montana to Santa Fe, where they buy their hobby ranches, ski-town condos, adobe villas, and settle in, telling us how much they love the West. But will not lift a finger to help defend it. Dante had a special place for these ESTers, esthetes, temporizers, and castrate fence-straddlers. He locked them in the vestibule of Hell. They're worse than the simple industrial developer, whose only objective, while pretending to "create jobs," is to create for himself a fortune in paper money. The developer is what he is: no further punishment is necessary. . . .

Maybe we should everyone stay home for a season, give our little Western wilderness some relief from Vibram soles, rubber boats. . . . But where is home? Surely not the walled-in prison of the cities, under that low ceiling of carbon monoxide . . . For a mere 5,000 years we have grubbed in the soil and laid brick upon brick to build the cities: but for a million years before that we lived the leisurely, free, and adventurous life of hunters and gatherers, warriors and tamers of horses. . . .

Ah yes, you say, but what about Mozart? Punk rock? Astrophysics? Flush toilets? Potato chips? Silicon chips? Oral surgery? Our coming journey to the stars? Vital projects, I agree, and I support them. But why not a compromise? Why can't we have a moderate number of small cities, bright islands of electricity and *kultur* and industry surrounded by shoals of farmland, cow range, and timberland, set in the midst of a great unbounded sea of primitive forest, virgin desert? The human reason can conceive of such a free and spacious world. Why can't we allow it to become—again—our home?

The American Indians had no word for what we call "wilderness." For them the wilderness was home. . . .

There will always be a Grand Canyon. There will always be a Rio Dolores, dam or no dam. There will always be one more voyage down the river to Bedrock, Colorado, in that high lonesome valley the pioneers named Paradox. A paradox because—anomaly—the river flows across, not through, the valley, apparently violating both geo-logic and common sense. Not even a plateau could stop the river. Their dams will go down like dominoes. And another river be reborn.

There will always be one more river. The journey goes on forever on our little living ship of stone and soil and water and vapor, this delicate planet circling round the sun which humankind call Earth.

Joy shipmates joy.

Overpopulated America

Wayne H. Davis

I define as most seriously overpopulated that nation whose people by virtue of their numbers and activities are most rapidly decreasing the ability of the land to support human life. With our large population, our affluence, and our technological monstrosities, the United States wins first place by a substantial margin.

Let's compare the United States to India, for example. We have 203 million people, whereas she has 540 million on much less land. But look at the impact of people on the land.

The average Indian eats his daily few cups of rice (or perhaps wheat, whose production on American farms contributed to our 1 percent per year drain in quality of our active farmland), draws his bucket of water from the communal well, and sleeps in a mud hut. In his daily rounds to gather cow dung to burn to cook his rice and warm his feet, his footsteps, along with those of millions of his countrymen, help bring about a slow deterioration of the ability of the land to support people. His contribution to the destruction of the land is minimal.

An American, on the other hand, can be expected to destroy a piece of land on which he builds a home, garage, and driveway. He

The New Republic, January 10, 1970.

will contribute his share to the 142 million tons of smoke and fumes, seven million junked cars, 20 million tons of paper, 48 billion cans, and 26 billion bottles the overburdened environment must absorb each year. To run his air conditioner we will strip-mine a Kentucky hillside, push the dirt and slate down into the stream, and burn coal in a power generator, whose smokestack contributes to a plume of smoke massive enough to cause cloud seeding and premature precipitation from Gulf winds which should be irrigating the wheat farms of Minnesota.

In his lifetime he will personally pollute 3 million gallons of water, and industry and agriculture will use ten times this much water in his behalf. To provide these needs the U.S. Army Corps of Engineers will build dams and flood farmland. He will also use 21,000 gallons of leaded gasoline containing boron, drink 28,000 pounds of milk, and eat 10,000 pounds of meat. The latter is produced and squandered in a life pattern unknown to Asians. A steer on a Western range eats plants containing minerals necessary for plant life. Some of these are incorporated into the body of the steer which is later shipped for slaughter. After being eaten by man these nutrients are flushed down the toilet into the ocean or buried in the cemetery, the surface of which is cluttered with boulders called tombstones and has been removed from productivity. The result is a continual drain on the productivity of range land. Add to this the erosion of overgrazed lands, and the effects of the falling water table as we mine Pleistocene deposits of groundwater to irrigate to produce food for more people, and we can see why our land is dying far more rapidly than did the great civilizations of the Middle East, which experienced the same cycle. The average Indian citizen, whose fecal material goes back to the land, has but a minute fraction of the destructive effect on the land that the affluent American does.

Thus, I want to introduce a new term, which I suggest be used in future discussions of human population and ecology. We should speak of our numbers in "Indian equivalents." An Indian equivalent I define as the average number of Indians required to have the same detrimental effect on the land's ability to support human life as would the average American. This value is difficult to determine, but let's take an extremely conservative working figure of 25. To see how conservative this is, imagine the addition of 1,000 citizens to your town and 25,000 to an Indian village. Not only would the Americans destroy much more land for homes, highways, and a shopping center, but they would contribute far more to environmental deterioration in hundreds of other ways as well. For example, their demand for steel for new autos might increase the daily pollution equivalent of 130,000 junk autos which *Life* tells us that U.S. Steel Corporation dumps into Lake Michigan. Their demand for textiles would help the cotton industry destroy the life in the Black Warrior River in Alabama with endrin. And they would contribute to the massive industrial pollution of our oceans (we provide one-third to one-half the world's share) which has caused the precipitous downward trend in our commercial fisheries landings during the past seven years.

The per capita gross national product of the United States is 38 times that of India. Most of our goods and services contribute to the decline in the ability of the environment to support life. Thus it is clear that a figure of 25 for an Indian equivalent is conservative. It has been suggested to me that a more realistic figure would be 500.

In Indian equivalents, therefore, the population of the United States is at least 4 billion. And the rate of growth is even more alarming. We are growing at 1 percent per year, a rate which would double our numbers in seventy years. India is growing at 2.5 percent. Using the Indian equivalent of 25, our population growth becomes 10 times as serious as that of India. According to the Reinows in their recent book *Moment in the Sun*, just one year's crop of American babies can be expected to use up 25 billion pounds of beef, 200 million pounds of steel, and 9.1 billion gallons of gasoline during their collective lifetime. And the demands on water and land for our growing population are expected to be far greater than the supply available in the year 2000. We are destroying our land at a rate of over a million acres a year. We now have only 2.6 agricultural acres per person. By 1975 this will be cut to 2.2, the critical point for the maintenance of what we consider a decent diet, and by the year 2000 we might expect to have 1.2.

You might object that I am playing with statistics in using the Indian equivalent on the rate of growth. I am making the assumption that today's Indian child will live thirty-five years (the average Indian life span) at today's level of influence. If he lives an American seventy years, our rate of population growth would be 20 times as serious as India's.

But the assumption of continued affluence at today's level is unfounded. If our numbers continue to rise, our standard of living will fall so sharply that by the year 2000 any surviving Americans might consider today's average Asian to be well off. Our children's destructive effects on their environment will decline as they sink lower into poverty.

The United States is in serious economic trouble now. Nothing could be more misleading than today's affluence, which rests precariously on a crumbling foundation. Our productivity, which had been increasing steadily at about 3.2 percent a year since World War II, has been falling during 1969. Our export over import balance has been shrinking steadily from $7.1 billion in 1964 to $1.05 billion in the first half of 1969. Our balance of payments deficit for the second quarter was $3.7 billion, the largest in history. We are now importing iron ore, steel, oil, beef, textiles, cameras, radios, and hundreds of other things.

Our economy is based upon the Keynesian concept of a continued growth in population and productivity. It worked in an underpopulated nation with excess resources. It could continue to work only if the earth and its resources were expanding at an annual rate of 4 to 5 percent. Yet neither the number of cars, the economy, the human population, nor anything else can expand indefinitely at an exponential rate in a finite world. We must face this fact *now.* The crisis is here. When Walter Heller says that our economy will expand by 4 percent annually through the latter 1970s he is dreaming. He is in a theoretical world totally unaware of the realities of human ecology. If the economists do not wake up and devise a new system for us now somebody else will have to do it for them.

A civilization is comparable to a living organism. Its longevity is a function of its metabolism. The higher the metabolism (affluence), the shorter the life. Keynesian economics has allowed us an affluent but shortened life span. We have now run our course.

The tragedy facing the United States is even greater and more imminent than that descending upon the hungry nations. The Paddock brothers in their book, *Famine 1975!*, say that India "cannot be saved" no matter how much food we ship her. But India will be here after the United States is gone. Many millions will die in the most colossal famines India has ever known, but the land will survive and she will come back as she always has before. The United States, on the other hand, will be a desolate tangle of concrete and ticky-tacky, of strip-mined moonscape and silt-choked reservoirs. The land and water will be so contaminated with pesticides, herbicides, mercury fungicides, lead, baron, nickel, arsenic, and hundreds of other toxic substances, which have been approaching critical levels of concentration in our environment as a result of our numbers and affluence, that it may be unable to sustain human life.

Thus as the curtain gets ready to fall on man's civilization let it come as no surprise that it shall first fall on the United States. And let no one make the mistake of thinking we can save ourselves by "cleaning up the environment." Banning DDT is the equivalent of the physician's treating syphilis by putting a Band Aid over the first chancre to appear. In either case you can be sure that more serious and widespread trouble will soon appear unless the disease itself is treated. We cannot survive by planning to treat the symptoms such as air pollution, water pollution, soil erosion, etc.

What can we do to slow the rate of destruction of the United States as a land capable of supporting human life? There are two approaches. First, we must reverse the population growth. We have far more people than we can continue to support at anything near today's level of affluence. American women average slightly over three children each. According to the *Population Bulletin* if we reduced this number to 2.5 there would still be 330 million people in the nation at the end of the century. And even if we reduced this to 1.5 we would have 57 million more people in the year 2000 than we have now. With our present longevity patterns it would take more than thirty years for the population to peak even when reproducing at this rate, which would eventually give us a net decrease in numbers.

Do not make the mistake of thinking that technology will solve our population problem by producing a better contraceptive. Our problem now is that people want too many children. Surveys show the average number of children wanted by the American family is 3.3. There is little difference between the poor and the wealthy, black and white, Catholic and Protestant. Production of children at this rate during the next thirty years would be so catastrophic in effect on our resources and the viability of the nation as to be beyond my ability to contemplate. To prevent this trend we must not only make contraceptives and abortion readily available to everyone, but we must establish a system to put severe economic pressure on those who produce children and reward those who do not. This can be done within our system of taxes and welfare.

The other thing we must do is to pare down our Indian equivalents. Individuals in American society vary tremendously in Indian equivalents. If we plot Indian equivalents versus their reciprocal, the percentage of land surviving a generation, we obtain a linear regression. We can then place individuals and occupation types on this graph. At one end would be the starving blacks of Mississippi; they would approach unity in Indian equivalents, and would have the least destructive effect on the land. At the other end of the graph would be the politicians slicing pork for the barrel, the highway contractors, strip-mine operators, real estate developers, and public enemy number one—the U.S. Army Corps of Engineers.

We must halt land destruction. We must abandon the view of land and minerals as private property to be exploited in any way economically feasible for private financial gain. Land and minerals are resources upon which the very survival of the nation depends, and their use must be planned in the best interests of the people.

Rising expectations for the poor is a cruel joke foisted upon them by the Establishment. As our new economy of use-it-once-and-throw-it-away produces more and more products for the affluent, the share of our resources available for the poor declines. Blessed be the starving blacks of Mississippi with their outdoor privies, for they are ecologically sound, and they shall inherit a nation. Although I hope that we will help these unfortunate people attain a decent standard of living by diverting war efforts to fertility control and job training, our most urgent task to assure this nation's survival during the next decade is to stop the affluent destroyers.

Waste a Lot, Want a Lot

Stuart Ewen

Our All-Consuming Quest for Style

Each day of the year, New York City disposes of 14,329 tons of garbage. Miami generates 7,445 tons; Los Angeles, 6,193 tons; Chicago, 5,985 tons; and Dallas, 1,948 tons of garbage per day.

Much of what gets thrown away is packaging, the provocatively designed wrappings that we have come to expect on nearly everything we purchase. But it is not all packaging. Increasingly, products that in the past would have been considered durables quickly find their way into the trash bin. These include wristwatches, telephones and other electronic devices, razors, pens, medical and hospital supplies, cigarette lighters, and, recently, cameras. General Electric and GTE sell $25 lamps designed to be discarded when the bulbs

All-Consuming Images: The Politics of Style in Contemporary Culture (New York: Basic Books, 1988).

burn out. Black and Decker sells a "throwaway travel iron." Predictions abound that automobile engines will soon be made of plastic and will be "less expensive to replace than repair."

From a marketing point of view, disposability is the golden goose. It conflates the act of *using* with that of *using up*, and promotes markets that are continually hungry for more. Joseph Smith, a consumer psychologist, contends that the popular appeal of disposable products "reflects our changing social values; there's less emphasis on permanence today."

Along these lines, John Rader Platt, professor of physics at the University of Chicago, has raised the issue of the "fifth need of man": "The needs of man, if life is to survive, are usually said to be four—air, water, food, and in the severe climates, protection. But it is becoming clear today that the human organism has another absolute necessity. . . . This fifth need is the need for novelty—the need, throughout our waking life, for continuous variety in the external stimulation of our eyes, ears, sense organs, and all our nervous network."

The ever-mounting glut of waste materials is a characteristic by-product of modern "consumer society." It might even be argued that capitalism's continual need to find or generate markets means that disposability and waste have become the spine of the system. To *consume* means, literally, "to destroy or expend," and in the garbage crisis we confront the underlying truth of a society in which enormous productive capacities and market forces have harnessed human needs and desires, without regard to the long- or even short-term future of life on the planet.

If the course of consumerism is one of continuous exhaustion of resources, it must be acknowledged that for most people living within consumer society, waste is seen as an inherent part of the processes by which they obtain replenishment and pleasure. In societies where local production and subsistence agriculture provided people with most of their essential material needs, the resources that people relied upon and the need to use them carefully were obvious.

In large measure, a consumer society begins to erode this cycle. While providing many people with new standards of material abundance, the market in goods makes "where things come from" increasingly abstract and apparently insignificant. Simultaneously, modern systems of waste disposal tend to make "where things go" seemingly inconsequential. The often-cited difficulty that *moderns* have in dealing with the life-process issues of birth, aging, and death, may, to some extent, reflect this ignorance of natural processes. We buy our chicken, cut up in a plastic tray, in a modern supermarket with canned music piped in. After we are all done, we place the bones, along with the plastic tray, along with all other kinds of waste, into a large plastic garbage bag which we place outside to be carted off. On those occasions when we *do* pass a dump site, there is little sense of personal belongings. It does not occur to us that it is *our* refuse.

The customs and values of rural and preindustrial life preached against waste or obsolescence, both of which are hallmarks of modern commercial culture. The historical roots of our current wasteful consumer sensibility lie in the social development of industrial capitalism, and in the apparently inexhaustible capacities of mass production.

Historically, the majority of people on earth lived by economic assumptions that were starkly at odds with those that guide the present. The customary limits governing economics appeared, to many, to have been broken, beginning in the late nineteenth century. In addition, new techniques, materials, and colorful chemical dyes foretold a world of universal abundance, where all might come to enjoy goods and privileges customarily available only to elites.

Much of late nineteenth and twentieth century social thought is premised on the coming of what historian Warren Susman termed "a newly emerging culture of abundance." To some extent, socialist thought has been fertilized by this expectation. For Karl Marx, writing at mid-nineteenth century, "the rapid improvement of all instruments of production" under capitalism was creating the material conditions that ultimately made communism possible.

In 1892, when most socioeconomic thought still "defended the assumption of scarcity" as an irrevocable part of the human condition, American social theorist Simon Patten declared that "enough goods and services would be produced in the foreseeable future to provide every human being with the requisites of survival." Old practices of saving and preservation, Patten observed, were on the wane, as was the customary way of life that had nurtured those traditional values. Mass production had raised all kinds of previously unimaginable possibilities, and the new standard of living would be drawn from among its myriad creations, propelled by the continual consumption of goods. "The standard of life is determined, not so much by what a man has to enjoy, as by the rapidity with which he tires of the pleasure," Patten wrote. "To have a high standard means to enjoy a pleasure intensely and to tire of it quickly."

The most visible symbol of waste is seen in the continually changing styles and packaging that, in order to stimulate sales, affect nearly every commodity. Here the principle of waste is not embedded in any particular image, but rather in the incessant spectacle that accompanies the marketing of merchandise. To a large extent, the phenomenon of market-stimulated waste had its beginnings in the early decades of the twentieth century, when emerging American consumer industries and the advertising agencies they hired methodically attempted to promote sales by assailing the "customs of ages," which had encouraged thrift and the careful preservation of resources. By the early 1920s, the advertising industry had begun to publicly define itself as both "the destroyer and creator in the process of the ever-evolving new."

Perhaps more than any other person, it was the advertising man Earnest Elmo Calkins who raised the strategy of rapid, planned stylistic changes as an element of twentieth century American business thinking. Calkins took aim at "the Puritan tradition" within American industry, a tradition that had achieved wonders in the area of technological efficiency, yet had little appreciation of the aesthetic dimension. His model for the outmoded "Puritan" businessman was Henry Ford. While many saw Ford as the father of the modern age, Calkins focused on what he saw as Ford's retrograde tendencies.

In going after Ford, Calkins selected as his text a remark Ford had made: "that he would not give five cents for all the art the world had produced." "There is no doubt that Mr. Ford was sincere in what he said about art," Calkins wrote. "He believed that the homeliness of his car was one of its virtues. He correctly read the minds of his fellow citizens, who suspected that mere prettiness camouflaged the fact that sterner virtues were lacking. The Ford car was homely, but it did its work."

By the mid-1920s, however, Ford's puritanical commitment to "homeliness" could no longer command an increasingly competitive market. By adding "design and color to mechanical efficiency," Calkins noted, Ford's primary competitor in the production of inexpensive cars, Chevrolet, seized the largest share of the market.

Employing the story of Ford as an object lesson, Calkins began to articulate a business strategy that not only added the question of style to the business agenda, but recommended ongoing, methodical style change as the key to business prosperity. Writing in *Modern Publicity* in 1930, Calkins compressed his strategy into one extraordinarily direct paragraph: "The purpose is to make the customer discontented with his old type of fountain pen, kitchen utensil, bathroom, or motor car, because it is old-fashioned, out-of-date. The technical term for this idea is obsoletism. We no longer wait for things to wear out. We displace them with others that are not more effective but more attractive."

By the 1930s, as the Depression intensified the ferocity of corporate competition for sales, the ideas laid out by Calkins became rules of thumb. Writing in *Consumer Engineering* in 1932, Roy Sheldon and Egmont Arens contended that modern conditions had forced a reassessment of meanings: "The dictionary gives obsolescence as the process by which anything—a word, a style, a machine—becomes antiquated, outworn, old-fashioned, falls into disuse—ceases to be used. But this definition is itself rapidly becoming obsolete. It expresses a pre-war point of view, [it] is passive." For Sheldon and Arens, it was necessary now to conceive of "obsolescence as a positive force," a resource to be used to drive the market forward.

Sheldon and Arens were particularly aware and enthusiastic about the role of the mass media in this development: "Obsolescence . . . is seething through the life of the nation. Every day the latest fashions in clothes, in furniture, in automobiles, in coiffures is flashed on the screen before 16,000,000 intent watchers. Heralded in the newspapers, illustrated in the magazines, described over the radio, the latest wrinkle and the newest gadget are pushing and crowding into people's lives, not casually or with the leisurely pace of prewar days, but with the haste and bustle—and also the gaiety—characteristic of the modern American *tempo.*"

If, during the 1930s, the practice of obsolescence was part of a sometimes desperate attempt to build markets in a shrinking economy, the period following World War II saw obsolescence installed as a basic underpinning to the "populuxe" ideal of suburban prosperity. Articulating the boom mentality of the period, industrial designer J. Gordon Lippincott hailed obsolescence as a fundamental American birthright. Writing in 1947, Lippincott noted that "we have become so used to change that as a nation we take it for granted. The American consumer *expects* new and better products every year. . . . His acceptance of change toward better living is indeed the American's greatest asset. It is the prime mover of our national wealth."

Yet even amid this period of suburban boosterism, it was clear to Lippincott that citizen expectations may not be enough. Back in the 1930s, Sheldon and Arens had noted that many people were resistant to the economy of waste. "Scratch a consumer," they wrote, "and you find an opponent of consumptionism and a fear of the workings of progressive obsolescence." By the late 1940s, conditions had changed, but Lippincott still spoke of the need to combat thrift-oriented thought: "The major problem confronting us is how to *move this merchandise to the American consumer.* The major problem therefore is one of stimulating the urge to buy! . . . Our willingness to part with something *before* it is completely worn out is . . . truly an American habit, and it is soundly based on our economy of abundance. It must be further nurtured even though it is contrary to one of the oldest inbred laws of humanity—the law of thrift—of providing for the unknown and often-feared day of scarcity."

Here, in the clarity of Lippincott's words, we confront the inner logic of the spectacle of waste: the live-for-the-moment ideology that primes the market and avoids the question of the future.

During the 1950s, the appeal to what Lippincott described favorably as "mass buying-psychosis" accelerated as never before. The suburbs were a symbolic escape from the rhythms of industrial society; they also represented the elevation of planned obsolescence as an entire way of life. At the center of this *lifestyle* was the automobile, which, during the 1950s, became a public laboratory in waste. Perhaps more than anyone else, Harley Earl—who coined the phrase "dynamic obsolescence" to describe the design approach he innovated at General Motors' Styling Department—expressed big business' approach to style and market stimulation in the mid-1950s when he said, "Design these days *means taking a bigger step every year.* Our big job is to hasten obsolescence. In 1934 the average car ownership span was five years; now it is two years. When it is one year, we will have a perfect score."

Population and Development Misunderstood

Anne Ehrlich and Paul Ehrlich

The National Research Council (NRC), the principal operating agency of the National Academy of Sciences and the National Academy of Engineering, recently published a report entitled *Population Growth and Economic Development: Policy Questions* that dramatically illustrates the pitfalls of attempting to deal with the human predicament without understanding elementary ecology. The study appears to have been an attempt to rebut the retrograde position of the Reagan administration on demographic issues, citing some of the lunatic fringe literature that has supported that position, and coming to the unexceptional conclusion that "family planning programs can play a role in improving the lives of people in developing countries."

That is a far cry from the conclusion the report should have come to—that population control is essential, not only for developing countries, but for all nations, if they are to have a decent future. How could a report from such a distinguished source, and written by highly competent economists and demographers, miss the point

The Amicus Journal, Summer 1986.

so thoroughly? One does not have to search far for an answer. There were no ecologists, evolutionists, or earth scientists among the members of the working group that produced the report, or on the NRC committee on population that supervised it. This omission was roughly equivalent to a report on the spread and impact of AIDS being written by a committee composed of first-rate sociologists and economists but no virologists, immunologists, epidemiologists, or physicians.

The report makes a fundamental error. It shows no real awareness that economic and social systems must operate within constraints set by the laws of physics, chemistry, and biology. Any competent evaluation of population and development should *start* with a careful analysis of the impact of population growth on the resources and ecological systems (ecosystems) of the nations concerned. It has long been clear from such analyses that Earth as a whole is overpopulated—by the simple standard that it is only capable, with today's technologies and social systems, of supporting roughly 5 billion people (many of them none too well) through the destruction and dispersal of its physical capital. That capital is a one-time bonanza produced over billions of years by natural processes, a bonanza that *Homo sapiens* has been the first species ever to exploit on a large scale.

The most obvious components of that capital are the classic nonrenewable resources: fossil fuels and other minerals. But more important in evaluating the human predicament today are soils, fresh water, and species (and their component genetically distinct populations) of other organisms. Soils and fresh water usually are regarded as renewable, but they are increasingly being converted into nonrenewable resources by human abuse of the environment. Soils that were created at rates of inches per millennium are being eroded away at rates of inches per decade. Agricultural economist Lester Brown evaluated the potential consequences of that difference in rates succinctly: "Civilization can survive the exhaustion of oil reserves, but not the continuing wholesale loss of topsoil."

Fossil groundwater supplies in many regions of the world, including the United States, are being pumped out at rates far exceeding those of recharge. The result sometimes is the collapse of underground aquifers or their invasion by salt water (destroying the resource). In some regions, surface recharge areas are being paved over, so, instead of percolating into the ground to replenish aquifers, rainwater runs off into the sea. Many aquifers are also being essentially permanently polluted by plumes of toxic wastes. The problems with groundwater supplies are far from trivial. For example, about half of all American citizens depend on wells for their drinking water, and roughly fifty times as much fresh water underlies the United States as falls on it annually as precipitation. In some countries, the ratio is even higher. And many developing countries are both water-short and increasingly dependent on groundwater.

Biological diversity is also a nonrenewable resource from the viewpoint of civilization; tens of millions of years are required for evolution to regenerate an equivalent array of populations and species after a serious episode of extinctions. All organisms are working parts of natural ecosystems, which support civilization by providing numerous irreplaceable services. These ecosystem services include controlling the quality of the atmosphere, ameliorating the weather, helping to run the hydrologic cycle (which supplies humanity with fresh water), generating and replenishing soils, disposing of wastes, recycling of nutrients essential to agriculture and forestry, pollinating many crops, controlling the vast majority of potential pests and carriers of human diseases, providing food from the sea, and maintaining a vast genetic library, from which humanity has already withdrawn the very basis of its civilization in the form of plants and animals to domesticate.

Biological diversity is now threatened with an episode of extinctions unparalleled since the catastrophe that wiped out the dinosaurs and many other groups of organisms some 65 million years ago. *Homo sapiens* is the source of the threat. Our species now, in one way or another, coopts and wastes some 40 percent of the earth's terrestrial food supply—a supply that must be shared among some five to thirty million species of animals. In the course of activities that degrade the environment, the expanding human population is attacking the very systems that provide its only significant income: the solar energy that green plants bind and make available in the process of photosynthesis. So while humanity is expending its capital, it is also destroying its source of income—in the process lowering the long-term carrying capacity of Earth for human life.

None of this is reflected in the NRC report. That it would not be was abundantly clear by its second page. There we read that "the earth's resources are finite, and more people by definition means fewer natural resources per person." So far, so good. But then comes: "Of course, the most important resources are not natural, but artificial (plants and equipment used in production, openings in school systems, jobs, social institutions, and economic infrastructure) and so are expandable."

It would be difficult to construct a clearer statement of the most elementary fallacy that infects mainline economics—that somehow the economic systems can operate without regard for the constraints imposed by the laws of physics, chemistry, and biology. This statement was, by the way, defended by one of the report's authors, a world-class social scientist, on the basis that people are willing to pay more for manufactured goods than natural resources! The old criticism of economics, that it can determine the price of everything and the value of nothing, clearly remains valid today.

There are other errors that betray the absence of knowledgeable natural scientists from the group preparing the study. The effects of acid rain are said to be reversible, although some of them (including, perhaps, many of the most severe ones in forests) are clearly irreversible on the time scale of interest to society. A confused chapter from a book which errs by several orders of magnitude on the energy collection efficiency of biomass and understates the amount of available sunlight by a factor of about two is cited as being "skeptical" of the utility of solar energy. The analysis in another article, which is cited as the basis for the NRC report's comments on the impact of population growth on environmental quality, was shown to give gross overestimates of the impact of population growth fifteen years ago in a paper published in *Science* magazine. As a final horrible example, the report states that the "issues related to investment in soil conservation are no different in principle from those related to other forms of investment. . . ." How many other forms of investment involve investing in something one's life depends upon, and which, if destroyed, cannot be replaced?

Errors such as these could and should be avoided in future NRC investigations of the human predicament. All that would be required would be for NRC committees investigating aspects of the human predicament to include not only appropriate social scientists but ecologists and other environmental scientists who are capable of placing the operation of social and economic systems in the crucial context.

But there is a more fundamental and potentially more beneficial solution. Ecologists are all trained in demography, since the dynamics of populations of other organisms follow the same rules of arithmetic as do human populations. And ecologists who are concerned with the relationship of the human population to its environment wrestle regularly with rather technical discourses in economics (small wonder, since both words, economics and ecology, come from the Greek root for house, and refer to human and natural housekeeping, respectively).

In contrast, many social scientists, including most economists, remain largely ignorant of the natural sciences in general and ecology in particular. One does not have to search far for the reason in economics—a perusal of the standard economics texts and graduate curricula shows no place where students of the dismal science could learn how the world works. Most economists studiously ignore the writing of pioneers in their discipline, such as Herman Daly of Louisiana State University, who have taken the time to inform themselves and now are trying to design sustainable economic systems.

It is high time that social scientists sought at least minimal training in environmental sciences, especially ecology. Ecology is a complex subject, but so are those that are normally explored by economists, demographers, sociologists, and political scientists. Some knowledge of ecology could improve enormously the social scientists' analyses of the human predicament, just as some knowledge of social science is crucial to ecologists who wish to address the topic. In both cases, familiarity with the basic tenets of the others' disciplines would improve communication between social and natural scientists and would greatly enhance the contributions of each toward solving the most important set of problems ever to confront humanity.

PART THREE

Toward Holism and Sustainability

To live is not enough; we must take part.

—Pablo Casals

"In a dark time," states Theodore Roethke, "the eye begins to see." And so it is that there are some who are seeing ways out of the morass. It is important that those who feel paralyzed by the spectacle of destruction within a system out of rational control know that others are witnessing the same destruction and are reacting to it in a variety of constructive ways. To see people taking positive action should help us to realize our own roles in effecting necessary change in a society that is laying waste to the planet.

The facets of a now perceptible move toward an ecological era are, for the most part, small and in poor communication with one another. As Thomas Berry wrote in 1988, "We are like a musician who faintly hears a melody deep within the mind, but not clearly enough to play it through."

The melody continues to grow, though, both in strength and in clarity, and as this takes place assumptions that not long ago seemed invincible are crumbling at the edges. Everywhere there are local grass-roots movements that have in common a strong sense of human-earth connection. There is rising a new kind of thinking that draws heavily on spiritual insights associated with Eastern philosophy, and now, increasingly, on Native American values and concepts. "We Indians will show this country how to act human," Vine Deloria, Jr., wrote twenty years ago. "Someday," he promised, "this country will revise its constitution, its laws, in terms of human beings, instead of property."

The first two pieces in this section are by Native Americans. Jose Barreiro teaches that the traditions of indigenous peoples are rich in ecological concepts that can guide us from our destructive path. Oren Lyons, an Iroquois statesman, offers the philosophy that our actions should be judged by the effects to be felt by the seventh generation. How different that long-range view is from corporate society's preoccupation with quarterly profits. There follow two essays by ecofeminists: Carolyn Merchant looks at environmental restoration, while Judith Plant discusses bioregionalism. Then, Dave Foreman, one of the country's most prominent environmentalists, gives an account of the conditions which led him and his colleagues from moderate to radical environmentalism. And Herman Daly, a World Bank economist, discusses the possibility of a steady state economy while lambasting Wall Street imagery.

There are many who believe that significant environmental change will have to come from grass-roots movements rather than from the top. This view is presented by Alan Durning, and is followed by Larry Borowsky's essay on Cliff Humphrey, a "consummate" grass-roots activist. This book closes with a short piece by Thomas Berry, whose primary message is that spiritual connections with Nature are essential. It is Berry's conviction that we are preparing to return to "the primordial community of the universe, the earth, and all living things." His smooth prose and gentle optimism are a fitting termination for the section and for the anthology.

Before long, it seems, there will be reached the social "critical mass" necessary for a revolutionary change in direction. May it come quickly; a change of revolutionary nature is long overdue.

—BW

Indigenous Peoples Are the "Miners' Canary" of the Human Family

Jose Barreiro

At this crucial mid-point in the journey from Earth Day 1970—the first public, global recognition of the Natural World's true relationship to human society—to the turn of the second millennium, I would remind all interested peoples to once again consider the centrality of the Indigenous Peoples' quest for justice to the survival of the planet.

The Indigenous Peoples are the "miners' canary" of the human family. In some cases, such as the meso American and Amazonian rain forests, they are the only cultures that understand the relationship of human society to those fragile ecosystems. But as heretofore unexplored habitats begin to be destroyed, we are forced to witness the final genocide of Indigenous Peoples the world over.

In many respects, to support the survival struggles of the Amazo-

Environmental Action, "Visions," May/June 1985.

nian Indians is to support the protection of the rain forests, is to support the breathing capacity of the Earth itself.

In the Northern Hemisphere, Indian community after Indian community is beset by groundwater contamination (Akwesasne), by huge hydroelectric dams (Northern Dene), by uranium mining and milling (Southwest Navajos and Pueblos). Conversely, at this crucial moment in history, an Indigenous-based intelligentsia is arising, and community-based development—focused on renewable sources of energy and local self-reliance—is regaining momentum among Native Peoples.

Internationally, the conditions and messages of the Indigenous Peoples are approaching a permanent forum at the United Nations. As Iroquois statesman Oren Lyons has pointed out, "The day must come when we see at the United Nations a seat for the eagle, a recognition of the whale, a love and respect for our own mother, the Earth."

Indigenous Peoples deserve recognition among the networks and nations of the world that they are the Elders of the Human Family.

Indigenous cultures are rich in ecological concept. "Our Mother the Earth" is a reality in the cosmologies of virtually every native people in the world. This is a recognition that does not die in the Indian Country. It comes from the old religions and even where the ancient lifeways have changed, there are people who keep this aspect of the culture alive. It has survived, against overwhelming brutality and stupidity, because it is a spiritual, generational belief. It is one of the currents of thought that make up Pan-Indigenous philosophy and a basic message of the Indian peoples.

The Indigenous oral traditions are wealthy in ecological concept. Consider the "Circle concept," and the high place "reciprocity" receives in the reflections of Indigenous Holy or Medicine or Religious people. Many of the laws governing the how and why of things returning to their source are universal to native cosmologies. There is the principle of the Warning of Purification, which takes on the magnitude of prophesy. Many oral traditions point to a time when the Earth's waters and climates would go askew, when new "diseases" will spring up, when the Indian message will raise itself up, to an international audience. Much of this has come to pass.

Earth Day 1970 represented a momentous breakthrough in the wall of cultural myopia that the Western World had put between itself and the Natural World. But it is also as though the rest of humanity arrived at the cosmological home of the Indigenous Peoples and, once there, failed to greet the host family properly.

It is time that proper alignments were made; it is time to support the Elders of the Human Family. Ecologically concerned people the world over should consider this multi-cultural reality.

There are oral messages of the deepest meaning coming from these elder cultures, and they are as spiritually sound as they are pragmatically accurate. They warn ominously of humankind's current collision course with the life-nurturing powers of our Mother, our Earth.

An Iroquois Perspective

Oren Lyons

A large perspective is required to understand the prospects of very real situations concerning the environment, of natural resources, of land claims, of interactions among nations, and of the welfare of not only mankind but the welfare of the natural world and all life. I will give you our perspective and you may agree or disagree.

There exists just to the south of Syracuse, New York, the central council of the chiefs of the Onondaga nation, part of the Six Nations of the Iroquois Confederacy. Onondaga is also the capital of the Six Nations. Meetings there of the Onondaga nation, and of the Six Nations, are an uninterrupted continuation of a government that's perhaps a thousand years old. The basis of our nation is that the sovereignty of the individual is supreme. We hear the word sovereignty bandied about and discussed and interpreted, but it is essentially a very simple and a very direct application of reality as we see it. Sovereignty is the act thereof. You are as sovereign as you are. An example of sovereignty was Idi Amin in Uganda. Regardless of the position that you may take about Amin, very recently in Uganda,

American Indian Environments: Historical, Legal, Ethical, and Contemporary Perspectives (Syracuse, N.Y.: Syracuse University Press, 1980).

Amin carried out sovereignty in a horrendous manner which destroyed thousands and thousands of lives. It was not a just sovereignty, but it was a very absolute form of sovereignty. And that is the point.

The action of a people in a territory, the ability and willingness of a people to defend that territory, and the recognition of that ability by other nations: that's a definition of the practical application of sovereignty. It's very simple. It has to deal with government, power and people, and force. The history of my people, of the Ho-de-no-sau-ne, is a long history which deals in the principles of peace: basically peace and the power to keep the peace. Peace, equality, and justice for people is given over into the hands of the chiefs, the welfare of all living things. In our perception all life is equal, and that includes the birds, animals, things that grow, things that swim. All life is equal in our perception. It is the Creator who presents the reality, and as you read this singularly, by yourself in your sovereignty and in your being and in your completeness, you are a manifestation of the creation. You are sovereign by the fact that you exist. And in this, the relationship demands respect for the equality of life. These are the principles through which the council governs in their sense of duty. We are a government that is intertwined with spiritual guidance. The first duty of the chiefs is to see that we conduct ceremonies precisely. That is the first duty. Only after that do we sit in council for the welfare of our people. So you can see the separation of spiritual, religious ways from political ways does not exist within the structure of the Ho-de-no-sau-ne, and also I might add, to most of all the other Indian nations as they had previously existed. There has been a great change in the affairs of our people, the manners of government.

We have sat through, as one of our elders said, five days of invasion, five days that our white brothers have been here, and in those five days, there has been tremendous change. We are looking ahead, as is one of the first mandates given to us as chiefs, to make sure and to make every decision that we make relate to the welfare and well-being of the seventh generation to come, and that is the basis by which we make decisions in council. We consider: will this be to the benefit of the seventh generation? That is a guideline. We have watched various forms of governments, we have watched internationally the development of industry. We have watched within our own nations and territories the exploitation of not only the people but the resources without regard to the seventh generation to come. We are facing together, you and I, your people and my people, your children and my children, we are facing together a very bleak future. There seems to be at this point very little consideration, minimum consideration, for what is to occur, the exploitation of wealth, blood, and the guts of our mother, the earth: Without the earth, without your mother, you could not be sitting here; without the sun, you would not be here. Over a time, a period, I've seen people become more and more distant and unrelated. Nevertheless, these very basic elements vitally concern you. The reality that you sit in, you're surrounded with at this point, this room that separates you from the true reality, the true power, the true understanding, the manifestation of these walls can also be likened to the manifestation of books, encyclopedias, interpretations as when you heard laws. In one rationalization upon another, you continue the exploitation for wealth and power. But you must consider in the process and in choosing the direction of your life: how will this affect the seventh generation? The system that we have observed not only here but internationally demands exploitation; industrialization demands a power base. It demands work forces for the gods of profit. It is our observation that the cathedrals that you worship in are not the ones that ring your bells on Sunday. The cathedrals that you worship in are the shopping malls that are found in this country.

Respect the proper manner so that the seventh generation will have a place to live in. Let us look at the large issues. We are concerned with all the children of this earth. We are concerned with the four colors of Man. Natural Law is very simple. You cannot change it: it prevails over all. There is not a tight rule, there is not a court, there is not a group of nations in this world that can change this Natural Law. You are subject and born to those Natural Laws. The Indians understood the Natural Laws. They built their laws to coincide with the Natural Laws. And that's how we survived.

Will people of all races learn? Will they observe? Will they reach beyond feelings of racism and antagonism to see what is good for the welfare of all people? And not only people, but all things that live. The water is shallow. There is less of it. There are 4 billion people in the world today. In the year 2009 there are going to be 8 billion people on this earth. In thirty years' time, well within your lifetime, you are going to be faced with double the population—and double the problems. What about that seventh generation? Where are you taking them? What will they have?

You may not agree with what I said, but in the course of freedom—and recognizing the sovereignty of the individual and his ability to be free—that's your choice. Define for yourselves your directions. Think about it. Today belongs to us, tomorrow we'll give it to the children, but today is ours. You have the mandate, you have the responsibility. Take care of your people—not yourselves, your people.

Restoration and Reunion with Nature

Carolyn Merchant

Restoration is a backward-looking philosophy. But unlike romanticism, which is a longing for the past, or preservation, which seeks to save what already exists, restoration implies an active participation in bringing the past back to life. It recognizes that, while humans may be part of nature, they also have more power to alter it than do other species. Admitting this, it goes on to provide ways to use that power responsibly and ethically by going back in time to heal what has been changed or damaged. But this very act, even as in some ways it reaches into the past, also creates a new future.

In reconstructing natural ecosystems such as prairies, forests, rivers, and lakes, humans are imitators of nature. By studying and mimicking natural patterns they can recover not only the communities themselves but some of the wisdom inherent in both cultural and biological evolution. Rather than taking nature apart and simplifying ecosystems, as the past three centuries of mechanistic science have taught us to do supremely well, restorationists are actively

Restoration & Management Notes, Winter 1986.

putting them back together. Rather than analyzing nature for the sake of dominating and controlling it, restorationists are synthesizing it for the sake of living symbiotically within the whole.

Mimesis, the process of imitation through which restoration takes place, has had an important history. Indian hunters mimicked the sounds, smells, and behavior of the animals they captured for food. Forest clearings planted with corn, beans, and squash by Indian women mimicked nature's polycultural patterns. The Indian's oral-aural culture of myths, songs, and poems by which tribal values were preserved were grounded in the mimetic, oral mode of knowing. In peasant agriculture, peasants danced in the fields to awaken the generative powers of nature and spread cider, cake, or corn on the ground to influence the seasonal cycles. The alchemist who followed in nature's footsteps to imitate her ways was participating in the natural cycles in order to hasten them. The miner who cajoled nature through prayer before following a vein of ore, and the smith who abstained from drinking and eating before shaping a metal on the anvil, were artists uncovering nature's own hidden patterns.

Platonism in ancient Greece and mechanistic science in early modern Europe both undermined the mimetic tradition by elevating analysis to a position of reverence. To Plato, mimesis was simply a catalogue of responses learned by rote. The knower should be separated from the known; the subject from the object. Not recollection and participation, but problem-solving and analysis were what mattered most. The song and narrative were replaced by logic, arithmetic, and science. Two millennia later, Newtonian scientists undertook to understand nature by dividing it into atomic parts and changing it through external forces. To Francis Bacon, imitation meant obeying nature in order to command her. Nature was to be dominated, not by following but by prodding and ferreting out her secrets. The organic cosmos of Aristotle, in which nature acted and developed from within, gave way to a world view that sanctioned external manipulation and control. The model of the technician repairing the clock from the outside superseded that of the artist who revealed the form inherent in the matter, or the doctor, whose herbs healed the body because their inner knowledge (scientia) became one with the body's own knowledge.

Drawing on the mechanistic model, modern agriculture has increasingly moved in the direction of artificial ecosystems occupied by monocultures that are vulnerable to pest outbreaks and catastrophic collapse. Identical fields outlined in precise geometric patterns for efficient cultivation and harvesting replicate lattice-like atomic patterns, replacing the diversity of small, haphazard patchworks of fields created in forest clearings. Further stimulated by urbanization and industrialization, traditional agriculture was profoundly altered during the agricultural improvement movements of the eighteenth and nineteenth centuries by the introduction of more efficient machinery and irrigation technology and by improvements in crop and animal breeding, artificial fertilizers, and chemical pesticides. As a result, the external energy needed to produce the chemicals; operate the farm machinery; and process, store, and transport the products often surpasses the calories the foods themselves supply.

Today, restoration is part of a spectrum of emerging disciplines based on imitation, synthesis, and a creative reciprocity between humans and non-human nature. Both restoration and agroecology look back to traditional agriculture, combining it with ecology in order to design sustainable systems by mimicking nature. Together, these disciplines represent a spectrum of practices based on reestablishing contact with nature through imitation. Thus, much is being learned by studying the polycultural methods of traditional farmers, combining the wisdom of traditional agriculture developed over generations of trial and error with an understanding of local ecology. In the resulting agroecosystems, the spatial arrangements and seasonal development of wild plant species are used as models; the farmer imitates the arrangement of local species of grass, vine, shrub, and tree to design integrated cereal, vegetable, fruit, and tree crop systems.

Similarly, agroforestry restores complementary arrangements of trees, crops, and animals in accord with ecological principles in order to maintain productivity without environmental degradation. Orchards are planted with a ground cover of legumes or berries and foraged by poultry, pigs, and bees to keep down pests and produce well-mulched and manured soil.

Permaculture carries the process of imitation a step further. As an agriculture for the future, it imitates ecosystem evolution toward climax states by designing perennial plant and animal crop interactions. In contrast to monocultural agriculture, permaculture uses several stories of trees, shrubs, vines, and perennial ground crops to absorb more light and nutrients, increasing the total yield. Plants and animals coexist in separate niches that reduce competition and promote symbiosis among species. Complexity not only helps to ward off catastrophes but increases the variety of foods produced. External energy and physical labor decrease as perennials mature, so that energy needs are provided from within.

The biological control of insects also uses natural ecosystems as models. Uncultivated land surrounding fields harbors birds and insect enemies as well as pests. Hedges and flowers along roadsides are attractive to beneficial insects. Diversity in crops and surroundings and arrangements of beneficial plants mimic natural conditions making crops less visible to insect enemies and acting as barriers to pest dispersal. Thus, by imitating nature, agricultural systems can be designed both to suppress pests and maximize total yield.

As a form of agriculture, restoration, too, is based on the capacity of both humans and nature for action. While restoration is oriented toward the reconstruction of authentic replicas of natural habitats, agriculture traditionally aims at the production of food, clothing, and shelter. In either case, however, the principle of mimesis is important: people can use the environment to fulfill real needs, while non-human nature acts reciprocally as a partner. In this way nature is used with respect, not as something passive and manipulable as in the mechanistic model but as a partner that is active and alive.

At a deeper level, a number of scientists in the past few years have proposed alternatives to the mechanistic framework based on nature's inherent activity, self-organization, permeable boundaries, and resilience. These deep structural changes in science itself may be indicative of the emergence of a new paradigm compatible with the recognition that a global crisis exists in current patterns of resource use.

The Gaia hypothesis of British chemist James Lovelock proposes that the earth's biota as a whole maintain an optimal, life-supporting chemical composition within the atmosphere and oceans. Gaia, the name of the Greek earth goddess, is a metaphor for a self-regulating system that controls the functioning of the earth's chemical cycles.

The thermodynamics of Ilya Prigogine contrasts the equilibrium and near-equilibrium dynamics of the closed, isolated physical systems described by the mechanistic model with open biological and social systems in which matter and energy are constantly being exchanged with their surroundings. In a similar spirit, the new physics of David Bohm contrasts the older world picture of atomic fragmentation with a new philosophy of wholeness expressed in the unfolding and enfolding of moments within a "holomovement." Bohm's cosmology emphasizes the primacy of process rather than the domination of parts.

These new theoretical frameworks share with action-oriented disciplines, such as restoration and permaculture, a participatory form of consciousness rooted in ecology. In opposition to the subject/object, mind/body, and nature/culture dichotomies basic to mechanistic science, ecological consciousness recognizes mind and skin as permeable boundaries that integrate organism with environment, tacit knowing and learning through visceral imitation, and complexity and process as a merging of nature with culture. Opposed to the abstract concepts of a disembodied intellect imposed on agriculture is the embeddedness of design in gardens that mimic natural patterns. Humans are neither helpless victims nor arrogant dominators of nature but active participants in the destiny of the systems of which they are a part.

References

Altieri, Miguel. 1983. *Agroecology: The Scientific Basis of Alternative Agriculture*. Berkeley, Ca.: Division of Biological Control, University of California, Berkeley.

Berman, Morris. 1981. *The Reenchantment of the World*. Ithaca: Cornell University Press.

Bohm, David. 1980. *Wholeness and the Implicate Order*. Boston: Routledge and Kegan Paul.

Bookchin, Murray. 1982. *The Ecology of Freedom: The Emergence and Dissolution of Hierarchy.* Palo Alto: Cheshire Books.

Devall, Bill and George Sessions. 1985. *Deep Ecology: Living As If Nature Mattered.* Salt Lake City: Peregrine Smith Books.

Jantsch, Erich. 1980. *The Self-Organizing Universe.* New York: Pergamon.

Jordan, William R. III. 1983. "Thoughts on Looking Back," *Restoration & Management Notes*, 1, no. 3 (Winter):2.

———. 1983. "On Ecosystem Doctoring," *Restoration & Management Notes*, 1, no. 4 (Fall):2.

Lovelock, James. 1979. *Gaia: A New Look at Life on Earth.* Oxford: Oxford University Press.

Merchant, Carolyn. 1980. *The Death of Nature: Women, Ecology, and the Scientific Revolution.* San Francisco: Harper and Row.

Mollison, Bill. 1984. *Permaculture Two: Practical Design for Town and Country in Permanent Agriculture.* Maryborough, Australia: Dominion Press-Hedges & Bell.

Mollison, Bill and David Holmgren. 1984. *Permaculture One: A Perennial Agriculture for Human Settlements.* Maryborough, Australia: Dominion Press-Hedges & Bell.

Prigogine, Ilya and Isabelle Stengers. 1984. *Order Out of Chaos: Man's New Dialogue With Nature.* Toronto: Bantam Books.

Searching for Common Ground: Ecofeminism and Bioregionalism

Judith Plant

It is no accident that the concept of ecofeminism has emerged from the many tendencies within the movement for social change. Women and nature have had a long association throughout history and it is only now that the deepest meanings of this association are being understood. Just as ecologists have paid critical attention to the attitudes, social structures, and rationalizations that have allowed the rape of the earth, so have feminists dug deeply to understand why society has rendered them second class citizens, at best.

Both schools of thought are now converging with similar analyses. The difference is that ecologists are scientists, basing their views of the interconnectedness of all things on the intellect, whereas feminists cannot help but come from the school of experience and have sought intellectual frameworks in order to try to make sense of their experience of subjugation. The coming together of the two

The New Catalyst, Winter 1987/88.

gives us hope for an understanding of the world that has the potential to be rooted in "thinking feelingly".

ECOLOGY AND WOMEN

Ecology is the study of the interdependence and interconnectedness of all living systems. As ecologists look at the consequences of changes in the environment, they are compelled to be critical of society. Because the natural world has been thought of as a *resource*, it has been exploited without regard for the life that it supports. Social ecology seeks ways to harmonize human and non-human nature, exploring how humans can meet their requirements for life and still live in harmony with their environments.

Ecology teaches us that life is in a constant state of change, as species seek ways to fit in particular environments which are, in turn, being shaped by the diversity of life within and around them. Adaptation is a *process*. Ecology helps develop an awareness of the need to incorporate these organic facts into our most general views of the world—those views that shape the way humans will *be* in the world.

Within human society, the idea of hierarchy has been used to justify social domination, and has been projected onto nature, thereby establishing an attitude of controlling the natural world. The convergence of feminism with ecology is occurring because of an increasing awareness that there are, in fact, no hierarchies in nature. A belief in the virtues of diversity and non-hierarchical organization is shared by both views.

Women have long been associated with nature: metaphorically, as in "Mother Earth," as well as with the naming of hurricanes and other natural disasters! Our language says it all: a "virgin" forest is one awaiting exploitation, as yet untouched by man (sic). In society, too, women have been associated with the physical side of life. Our role has been "closer to nature," our "natural" work centered around human physical requirements: eating, sex, cleaning, the care of children and sick people. We have taken care of day-to-day life so that men have been able to go "out in the world," to create and enact methods of exploiting nature, including other human beings.

Historically, women have had no real power in the outside world, no place in decision-making and intellectual life. Today, however, ecology speaks for the earth, for the "other" in human/environmental relationships; and feminism speaks for the "other" in female/male relations. And ecofeminism, by speaking for *both* the original others, seeks to understand the interconnected roots of all domination, as well as ways to resist and change. The ecofeminist's task is one of developing the ability to take the place of the other, when considering the consequences of possible actions, and ensuring that we do not forget that we are all part of one another.

ECOFEMINISM: ITS VALUES AND DIMENSIONS

Why does patriarchal society want to forget its biological connections with nature? And why does it seek to gain control over life in the form of women, other peoples, or nature? And what, on earth, can we do about dismantling this process of domination? What kind of society could live in harmony with its environment? These questions form the basis of the ecofeminist perspective.

Before the world was mechanized and industrialized, the metaphor that explained self, society and the cosmos was the image of organism. This is not surprising, since most people were connected with the earth in their daily lives, living a subsistence existence. The earth was seen as female. And with two faces: one, the passive, nurturing mother; the other, wild and uncontrollable.

These images served as cultural constraints. The earth was seen to be alive, sensitive: it was considered unethical to do violence toward her. Who could conceive of killing a mother, or digging into her body for gold, or of mutilating her? But, as society began to shift from a subsistence economy to a market economy, as European cities grew and forested areas shrunk, and as the people moved away from the immediate, daily organic relationships which had once been their basis for survival, peoples' cultural values—and thus their stories—had to change. The image of earth as passive and gentle receded. The "wrath and fury" of nature, as woman, was the quality that now justified the new idea of "power over nature." With the new technology, man (sic) would be able to subdue her.

The organic metaphor that once explained everything was replaced by mechanical images. By the mid-seventeenth century, society had rationalized the separation of itself from nature. With nature "dead" in this view, exploitation was purely a mechanical function and it proceeded apace.

The new images were of controlling and dominating: having power over nature. Where the nurturing image had once been a cultural restraint, the new image of mastery allowed the clearing of forests and the damming and poisoning of rivers. And human culture which, in organic terms, should reflect the wide diversity in nature, has now been reduced to mono-culture, a simplification solely for the benefit of marketing.

Since the subjugation of women and nature is a social construction, not a biologically determined fact, our position of inferiority can be changed. At the same time as we're creating the female as an independent individual, we can be healing the mind/body split.

Life struggles in nature, such as the Stein Valley and the many less-publicized ones, become feminist issues within the ecofeminist perspective. Once we understand the historical connections between women and nature and their subsequent oppression, we cannot help but take a stand on the war against nature. By participating in these environmental standoffs against those who are assuming the right to control the natural world, we are helping to create an awareness of domination at all levels.

Ecofeminism gives women and men common ground. While women may have been associated with nature, they have been socialized to think in the same dualities as men have and we feel just as alienated as do our brothers. The social system isn't good for either of us! Yet, we *are* the social system. We need some common ground from which to be critically self-conscious, to enable us to recognize and affect the deep structure of our relations, with each other and with our environment.

In addition to participating in forms of resistance, such as nonviolent civil disobedience in support of environmental issues, we can also encourage, support, and develop—within our communities—a cultural life which celebrates the many differences in nature, and encourages thought on the consequences of our actions, in all our relations.

Bioregionalism, with its emphasis on distinct regional cultures and identities strongly attached to their natural environments, may well be the kind of framework within which the philosophy of ecofeminism could realize its full potential as part of a practical social movement.

BIOREGIONALISM: AN INTEGRATING IDEA

Bioregionalism means learning to become native to place, fitting ourselves to a particular place, not fitting a place to our predetermined tastes. It is living within the limits and the gifts provided by a place, creating a way of life that can be passed on to future generations. As Peter Berg and Raymond Dasmann have so eloquently stated, it "involves becoming native to a place through becoming aware of the particular ecological relationships that operate within and around it. It means understanding activities and evolving social behavior that will enrich the life of that place, restore its life-supporting systems, and establish an ecologically and socially sustainable pattern of existence within it. Simply stated it involves becoming fully alive in and with a place. It involves applying for membership in a biotic community and ceasing to be its exploiter."

Understanding the limitations of political change-revolution—bioregionalists are taking a broader view, considering change in evolutionary terms. Rather than winning or losing, or taking sides, as being the ultimate objective, *process* has come to be seen as key to our survival. *How* we go about making decisions and how we act them out are as important as *what* we are trying to decide or do.

In evolutionary terms, a species' adaptation must be sustainable if the species is to survive. How can humans meet their requirements and live healthy lives? What would an ecologically sustainable human culture be like? It is in dealing with these questions that the bioregional movement and the philosophy of ecofeminism are very much interconnected.

Human adaptation has to do with culture. What has happened with the rise of civilization, and most recently with the notion of mass culture, is that what could be called bioregionally adapted human groups, *no longer can exist*. It's difficult to imagine how society could be structured other than through centralized institutions that service the many. In our culture almost every city exists beyond its carrying capacity: diverse regions are being exhausted and ecologically devastated.

Becoming native to a place—learning to live in it on a sustainable basis over time—is not just a matter of appropriate technology, home-grown food, or even "reinhabiting" the city. It has very much to do with a shift in morality, in the attitudes and behaviors of human beings. With the help of feminism, women especially have learned an intimate lesson about the way power works. We have painfully seen that it is the same attitude which allows violence toward us that justifies the rape of the earth. Literally, the images are the same. We also know that we are just as capable, generally speaking, of enacting the same kind of behavior.

The ideas of bioregionalism are being practiced all over the world—just rarely referred to as such. The name gives us common ground, however, like ecofeminism. But bioregionalism gives us something to practice and together they could be seen to offer a praxis—that is, a way of living what we're thinking. Here we can begin to develop an effective method of sharing with our male friends the lessons we have learned about power, as well as our hopes and aspirations for an egalitarian society—a society which would be based on the full participation and involvement of women and men in the process of adaptation and thus in the maintenance of healthy ecosystems.

HOMING IN ON A NEW IMAGE

One of the key ideas of bioregionalism is the decentralization of power; moving further and further toward self-governing forms of social organization. The further we move in this direction, the closer we get to what has traditionally been thought of as "woman's sphere"—that is, home and its close surroundings. Ideally, the bioregional view values home above all else, because it is here where new values and behaviors are actually created. Here, alternatives can root and flourish and become deeply embedded in our way of being. This is not the same notion of home as the bungalow in the suburbs of western industrialized society! Rather, it is the place where we can learn the values of caring for and nurturing each other and our environments, and of paying attention to immediate human needs and feelings. It is a much broader term, reflecting the reality of human cultural requirements and our need to be sustainably adaptive within our non-human environments. The word ecology, in its very name, points us in this direction: *oikos*, the Greek root of "eco" means home.

The catch is that, in practice, home, with all its attendant roles, will not be anything different from what it has been throughout recent history *without* the enlightened perspective offered by feminism. Women's values, centered around life-giving, must be revalued, elevated from their once subordinate role. What women know from experience needs recognition and respect. We have had generations of experience in conciliation, dealing with interpersonal conflicts in daily domestic life. We know how to feel for others because we have practiced it.

At the same time, our work—tending to human physical requirements—has been undervalued. What has been considered material and physical has been thought to be "less than" the intellectual, the "outside" (of home) world. Women have been very much affected by this devaluation and this is reflected in our images of ourselves and our attitudes toward our work. Men, too, have been alienated from childcare and all the rest of daily domestic life which has a very nurturing effect on all who participate. Our society has devalued the source of its human-ness.

Home is the theatre of our human ecology, and it is where we can effectively think feelingly. Bioregionalism, essentially, is attempting to rebuild human and natural community. We know that it is non-adaptive to repeat the social organization which left women and children alone, at home, and men out in the world doing the "important" work. The *real* work is at home. It is not simply a question of fairness or equality, it is because, as a species, we have to actually work things out—just as it is in the so-called natural world—with all our relations. As part of this process, women and nature, indeed *humans* and nature, need a new image, as we mend our relations with each other and with the earth. Such an image will surely reflect what we are learning through the study of ecology, what we are coming to understand through feminism, and what we are experiencing by participating in the bioregional project.

From Confessions of an Eco-Warrior

Dave Foreman

Earth First!

If opposition is not enough, we must resist. And if resistance is not enough, then subvert.

—Edward Abbey

The early conservation movement in the United States was a child—and no bastard child—of the Establishment. The founders of the Sierra Club, the National Audubon Society, The Wilderness Society, and the wildlife conservation groups were, as a rule, pillars

Confessions of an Eco-Warrior (New York: Harmony Books, 1991).

of American society. They were an elite band—sportsmen of the Teddy Roosevelt variety, naturalists like John Burroughs, outdoorsmen in the mold of John Muir, pioneer foresters and ecologists on the order of Aldo Leopold, and wealthy social reformers like Gifford Pinchot and Robert Marshall. No anarchistic Luddites, these.

When the Sierra Club grew into the politically effective force that blocked Echo Park Dam in 1956 and got the Wilderness Act passed in 1964, its members (and members of like-minded organizations) were likely to be physicians, mathematicians, and nuclear physicists. To be sure, refugees from the mainstream joined the conservation outfits in the 1950s and 1960s, and David Brower, executive director of the Sierra Club during that period, and the man most responsible for the creation of the modern environmental movement, was beginning to ask serious questions about the assumptions and direction of industrial society by the time the Club's board of directors fired him in 1969. But it was not until Earth Day in 1970 that the environmental movement received its first real influx of antiestablishment radicals, as Vietnam War protestors found a new cause—the environment. Suddenly, beards appeared alongside crewcuts in conservation group meetings—and the rhetoric quickened.

The militancy was short-lived. Eco-anarchist groups like Black Mesa Defense, which provided a cutting edge for the movement, peaked at the United Nations' 1972 Stockholm Conference on the Human Environment, but then faded from the scene. Along with dozens of other products of the 1960s who went to work for conservation organizations in the early 1970s, I discovered that compromise seemed to work best. A suit and a tie gained access to regional heads of the U.S.D.A. Forest Service and to members of Congress. We learned to moderate our opinions along with our dress. We learned that extremists were ignored in the councils of government, that the way to get a senator to put his arm around your shoulders and drop a Wilderness bill in the hopper was to consider the conflicts—mining, timber, grazing—and pare back the proposal accordingly. *Of course* we were good, patriotic Americans. *Of course* we were concerned with the production of red meat, timber, and minerals. We tried to demonstrate that preserving wilderness did not conflict all that much with the gross national product, and that clean air actually helped the economy. We argued that we could have our booming industry and still not sink oil wells in pristine areas.

This moderate stance appeared to pay off when Jimmy Carter, the first president who was an avowed conservationist since Teddy Roosevelt, took the helm at the White House in 1977. Self-professed conservationists were given decisive positions in Carter's administration. Editorials proclaimed that environmentalism had been enshrined in the Establishment, that conservation was here to stay. A new ethic was at hand: Environmental Quality and Continued Economic Progress.

Yet, although we had access to and influence in high places, something seemed amiss. When the chips were down, conservation still lost out to industry. But these were our friends turning us down. We tried to understand the problems they faced in the real political world. We gave them the benefit of the doubt. We failed to sue when we should have. . . .

I wondered about all this on a gray day in January 1979 as I sat in my office in the headquarters of The Wilderness Society, only three blocks from the White House. I had just returned from a news conference at the South Agriculture Building, where the Forest Service had announced a disappointing decision on the second Roadless Area Review and Evaluation, a twenty-month exercise by the Forest Service to determine which National Forest lands should be protected in their natural condition. As I loosened my tie, propped my cowboy boots up on my desk, and popped the top on another Stroh's, I thought about RARE II and why it had gone so wrong. Jimmy Carter was supposedly a great friend of wilderness. Dr. M. Rupert Cutler, a former assistant executive director of The Wilderness Society, was Assistant Secretary of Agriculture over the Forest Service and had conceived the RARE II program. But we had lost to the timber, mining, and cattle interests on every point. Of 80 million acres still roadless and undeveloped in the 190 million acres of National Forests, the Department of Agriculture recommended that only 15 million receive protection from road building and timber cutting.* Moreover, damn it, we—the conservationists—had been moderate. The antienvironmental side had been extreme, radical, emotional, their arguments full of holes. We had been factual, rational. We had provided more—and better—serious public comment. But we had lost, and now we were worried that some local wilderness group might go off the reservation and sue the Forest Service over the clearly inadequate environmental impact statement for RARE II. We didn't want a lawsuit because we knew we could win and were afraid of the political consequences of such a victory. We might make some powerful senators and representatives angry. So those of us in Washington were plotting how to keep the grassroots in line. Something about all this seemed wrong to me.

After RARE II, I left my position as issues coordinator for The Wilderness Society in Washington to return to New Mexico and my old job as the Society's Southwest representative. I was particularly concerned with overgrazing on the 180 million acres of public lands in the West managed by the Department of the Interior's Bureau of Land Management. For years, these lands—rich in wildlife, scenery, recreation, and wilderness—had been the private preserve of stock growers in the West. BLM had done little to manage the public lands or to control the serious overgrazing that was sending millions of tons of topsoil down the Colorado, the Rio Grande, and other rivers, wiping out wildlife habitat, and generally beating the land to hell.

Prodded by a Natural Resources Defense Council lawsuit, BLM began to address the overgrazing problem through a series of environmental impact statements. These confirmed that most BLM lands were seriously overgrazed, and recommended cuts in livestock numbers. But after the expected outcry from ranchers and their political cronies in Congress and in state capitals, BLM backtracked so quickly that the Department of the Interior building suffered structural damage. Why were BLM and the Department of the Interior so gutless?

* Only 62 million acres were actually considered by the Forest Service in RARE II. Another 18 million acres that were also roadless and undeveloped were not considered because of sloppy inventory procedures, political pressure, or because areas had already gone through land-use plans that had supposedly considered their wilderness potential.

While that question gnawed at my innards, I was growing increasingly disturbed about trends in the conservation organizations themselves. When I originally went to work for The Wilderness Society in 1973, the way to get a job with a conservation group was to prove yourself first as a volunteer. It helped to have the right academic background, but experience as a capable grassroots conservation activist was more important.

We realized that we would not receive the salaries we could earn in government or private industry, but we didn't expect them. We were working for nonprofit groups funded by the contributions of concerned people. Give us enough to keep food on the table, pay rent, buy a six-pack—we didn't want to get rich. But a change occurred after the mid-1970s: people seeking to work for conservation groups were career-oriented; they had relevant degrees (science or law, not history or English); they saw jobs in environmental organizations in the same light as jobs in government or industry. One was a steppingstone to another, more powerful position later on. They were less part of a cause than members of a profession.

A gulf began to grow between staff and volunteers. We also began to squabble over salaries. We were no longer content with subsistence wages, and the figures on our paychecks came to mark our status in the movement. Perrier and Brie replaced Bud and beans at gatherings.

Within The Wilderness Society, executive director Celia Hunter, a prominent Alaskan conservationist and outfitter, World War II pilot, and feminist, was replaced in 1978 by Bill Turnage, an eager young businessman who had made his mark by marketing Ansel Adams. Within two years Turnage had replaced virtually all those on the staff under Celia with professional organization people. The governing council also worked to bring millionaires with a vague environmental interest on board. We were, it seemed to some of us, becoming indistinguishable from those we were ostensibly fighting.

I resigned my position in June 1980.

The same dynamics seemed to affect the rest of the movement. Were there any radicals anywhere? Anyone to take the hard stands? Sadly, no. The national groups—Sierra Club, Friends of the Earth, National Audubon Society, Natural Resources Defense Council, and the rest—took almost identical middle-of-the-road positions on most issues. And then those half-a-loaf demands were readily compromised further. The staffs of these groups fretted about keeping local conservationists (and some of their field representatives) in line, keeping them from becoming extreme or unreasonable, keeping them from blowing moderate national strategy. Even Friends of the Earth, which had started out radical back in the heady Earth Day era, had gravitated to the center and, as a rule, was a comfortable member of the informal coalition of big environmental organizations.

For years I advocated this approach. We could, I believed, gain more wilderness by taking a moderate tack. We would stir up less opposition by keeping a low profile. We could inculcate conservation in the Establishment by using rational economic arguments. We needed to present a solid front.

A major crack in my moderate ideas appeared early in 1979, when I returned from Washington to the small ranching community of Glenwood, New Mexico. I had lived there earlier for six years, and although I was a known conservationist, I was fairly well accepted. Shortly after my return, *The New York Times* published an article on RARE II, with the Gila National Forest around Glenwood as chief exhibit. To my amazement, the article quoted a rancher who I considered to be a friend as threatening *my* life because of local fears about the consequences of wilderness designations. A couple of days later I was accosted on the street by four men, one of whom ran the town café where I had eaten many a chicken-fried steak. They threatened my life because of RARE II.

I was not afraid, but I was irritated—and surprised. I had been a leading moderate among New Mexico conservationists. I had successfully persuaded New Mexico conservation groups to propose fewer RARE II areas on the Gila National Forest as Wilderness. What had backfired? I thought again about the different approaches to RARE II: the moderate, subdued one advanced by the major conservation groups; the howling, impassioned, extreme stand set forth by off-road-vehicle zealots, many ranchers, local boosters, loggers, and miners. They looked like fools. We looked like statesmen. They won.

The last straw fell on the Fourth of July, 1980, in Moab, Utah.

There the local country commission sent a flag-flying bulldozer into an area the Bureau of Land Management had identified as a possible study area for Wilderness designation. The bulldozer incursion was an opening salvo for the so-called Sagebrush Rebellion, a move by chambers of commerce, ranchers, and right-wing fanatics in the West to claim federal public land for the states and eventual transfer to private hands. The Rebellion was clearly an extremist effort, lacking the support of even many conservative members of Congress in the West, yet BLM was afraid to stop the county commission.

What have we really accomplished? I thought. *Are we any better off as far as saving the Earth now than we were ten years ago?* I ticked off the real problems: world population growth, destruction of tropical forests, expanding slaughter of African wildlife, oil pollution of the oceans, acid rain, carbon dioxide buildup in the atmosphere, spreading deserts on every continent, destruction of native peoples and the imposition of a single culture (European) on the entire world, plans to carve up Antarctica, planned deep seabed mining, nuclear proliferation, recombinant DNA research, toxic wastes. . . . It was staggering. And I feared we had done nothing to reverse the tide. Indeed, it had accelerated.

And then: Ronald Reagan. James "Rape 'n' Ruin" Watt became Secretary of the Interior. The Forest Service was Louisiana-Pacific's. Interior was Exxon's. The Environmental Protection Agency was Dow's. Quickly, the Reagan administration and the Republican Senate spoke of gutting the already gutless Alaska Lands bill. The Clean Air Act, up for renewal, faced a government more interested in corporate black ink than human black lungs. The lands of the Bureau of Land Management appeared to the Interior Department obscenely naked without the garb of oil wells. Concurrently, the Agriculture Department directed the Forest Service to rid the National Forests of decadent and diseased old-growth trees. The cowboys had the grazing lands, and God help the hiker, Coyote, or blade of grass that got in their way.

Maybe, some of us began to feel, even before Reagan's election, it was time for a new joker in the deck: a militant, uncompromising group unafraid to say what needed to be said or to back it up with stronger actions than the established organizations were willing to take. This idea had been kicking around for a couple of years. Finally, in 1980, several disgruntled conservationists—including Susan Morgan, formerly educational director for The Wilderness Society; Howie Wolke, former Wyoming representative for Friends of the Earth; Bart Koehler, former Wyoming representative for The Wilderness Society; Ron Kezar, a longtime Sierra Club activist; and I—decided that the time for talk was past. We formed a new national group, which we called Earth First! We set out to be radical in style, positions, philosophy, and organization in order to be effective and to avoid the pitfalls of co-option and moderation that we had already experienced.

What, we asked ourselves as we sat around a campfire in the Wyoming mountains, were the reasons and purposes for environmental radicalism?

- To state honestly the views held by many conservationists.
- To demonstrate that the Sierra Club and its allies were raging moderates, believers in the system, and to refute the Reagan/Watt contention that they were "environmental extremists."
- To balance such antienvironmental radicals as the Grand County commission and provide a broader spectrum of viewpoints.
- To return vigor, joy, and enthusiasm to the tired, unimaginative environmental movement.
- To keep the established groups honest. By stating a pure, no-compromise, pro-Earth position, we felt that Earth First! could help keep the other groups from straying too far from their original philosophical base.
- To give an outlet to many hard-line conservationists who were no longer active because of disenchantment with compromise politics and the co-option of environmental organizations.
- To provide a productive fringe, since ideas, creativity, and energy tend to spring up on the edge and later spread into the center.
- To inspire others to carry out activities straight from the pages of *The Monkey Wrench Gang* (a novel of environmental sabotage by Edward Abbey), even though Earth First!, we agreed, would itself be ostensibly law-abiding.

• To help develop a new worldview, a biocentric paradigm, an Earth philosophy. To fight, with uncompromising passion, for Earth.

The name Earth First! was chosen because it succinctly summed up the one thing on which we could all agree: That in *any* decision, consideration for the health of the Earth must come first.

In a true Earth-radical group, concern for wilderness preservation must be the keystone. The idea of wilderness, after all, is the most radical in human thought—more radical than Paine, than Marx, than Mao. Wilderness says: Human beings are not paramount, Earth is not for *Homo sapiens* alone, human life is but one life form on the planet and has no right to take exclusive possession. Yes, wilderness for its own sake, without any need to justify it for human benefit. Wilderness for wilderness. For bears and whales and titmice and rattlesnakes and stink bugs. And . . . wilderness for human beings. Because it is the laboratory of human evolution, and because it is home.

It is not enough to protect our few remaining bits of wilderness. The only hope for Earth (including humanity) is to withdraw huge areas as inviolate natural sanctuaries from the depredations of modern industry and technology. Keep Cleveland, Los Angeles. Contain them. Try to make them habitable. But identify big areas that can be restored to a semblance of natural conditions, reintroduce the Grizzly Bear and wolf and prairie grasses, and declare them off limits to modern civilization.

In the United States, pick an area for each of our major ecosystems and recreate the American wilderness—not in little pieces of a thousand acres, but in chunks of a million or ten million. Move out the people and cars. Reclaim the roads and plowed land. It is not enough any longer to say no more dams on our wild rivers. We must begin tearing down some dams already built—beginning with Glen Canyon on the Colorado River in Arizona, Tellico in Tennessee, Hetch Hetchy and New Melones in California—and freeing shackled rivers.

This emphasis on wilderness does not require ignoring other environmental issues or abandoning social issues. In the United

States, blacks and Chicanos of the inner cities are the ones most affected by air and water pollution, the ones most trapped by the unnatural confines of urbanity. So we decided that not only should eco-militants be concerned with these human environmental problems, we should also make common ground with other progressive elements of society whenever possible.

Obviously, for a group more committed to Gila Monsters and Mountain Lions than to people, there will not be a total alliance with other social movements. But there are issues in which Earth radicals can cooperate with feminist, Native American, anti-nuke, peace, civil-rights, and civil-liberties groups. The inherent conservatism of the conservation community has made it wary of snuggling too close to these leftist organizations. We hoped to pave the way for better cooperation from the entire conservation movement.

We believed that new tactics were needed—something more than commenting on dreary environmental-impact statements and writing letters to members of Congress. Politics in the streets. Civil disobedience. Media stunts. Holding the villains up to ridicule. Using music to charge the cause.

Action is the key. Action is more important than philosophical hairsplitting or endless refining of dogma (for which radicals are so well known). Let our actions set the finer points of our philosophy. And let us recognize that diversity is not only the spice of life, but also the strength. All that would be required to join Earth First!, we decided, was a belief in Earth first. Apart from that, Earth First! would be big enough to contain street poets and cowboy bar bouncers, agnostics and pagans, vegetarians and raw-steak eaters, pacifists and those who think that turning the other cheek is a good way to get a sore face.

Radicals frequently verge on a righteous seriousness. But we felt that if we couldn't laugh at ourselves we would be merely another bunch of dangerous fanatics who should be locked up—like oil company executives. Not only does humor preserve individual and group sanity; it retards hubris, a major cause of environmental rape, and it is also an effective weapon. Fire, passion, courage, and emotionalism are also needed. We have been too reasonable, too calm, too understanding. It's time to get angry, to cry, to let rage flow at what the human cancer is doing to Earth, to be uncompromising. For Earth First! there is no truce or cease-fire. No surrender. No partitioning of the territory.

Ever since the Earth goddesses of ancient Greece were supplanted by the macho Olympians, repression of women and Earth has gone hand in hand with imperial organization. Earth First! decided to be nonorganizational: no officers, no bylaws or constitution, no incorporation, no tax status, just a collection of women and men committed to the Earth. At the turn of the century, William Graham Sumner wrote a famous essay titled "The Conquest of the United States by Spain." His thesis was that Spain had ultimately won the Spanish-American War because the United States took on the imperialism and totalitarianism of Spain. We felt that if we took on the organization of the industrial state, we would soon accept their anthropocentric paradigm, much as Audubon and the Sierra Club already had.

And when we are inspired, we *act*.

Massive, powerful, like some creation of Darth Vader, Glen Canyon Dam squats in the canyon of the Colorado River on the Arizona-Utah border and backs the cold, dead waters of "Lake" Powell some 180 miles upstream, drowning the most awesome and magical canyon on Earth. More than any other single entity, Glen Canyon Dam is the symbol of the destruction of wilderness, of the technological ravishment of the West. The finest fantasy of eco-warriors in the West is the destruction of the dam and the liberation of the Colorado. So it was only proper that on March 21, 1981—at the spring equinox, the traditional time of rebirth—Earth First! held its first national gathering at Glen Canyon Dam.

On that morning, seventy-five members of Earth First! lined the walkway of the Colorado River Bridge, 700 feet above the once-free river, and watched five compatriots busy at work with an awkward black bundle on the massive dam just upstream. Those on the bridge carried placards reading "Damn Watt, Not Rivers," "Free the Colorado," and "Let It Flow." The five of us on the dam attached ropes to a grille, shouted out "Earth First!" and let 300 feet of black plastic unfurl down the side of the dam, creating the impression of a growing crack. Those on the bridge returned the cheer.

Then Edward Abbey, author of *The Monkey Wrench Gang*, told the protestors of the "green and living wilderness" that was Glen Canyon only nineteen years ago:

> And they took it away from us. The politicians of Arizona, Utah, New Mexico, and Colorado, in cahoots with the land developers, city developers, industrial developers of the Southwest, stole this treasure from us in order to pursue and promote their crackpot ideology of growth, profit, and power—growth for the sake of power, power for the sake of growth.

Speaking toward the future, Abbey offered this advice: "Oppose. Oppose the destruction of our homeland by these alien forces from Houston, Tokyo, Manhattan, Washington, D.C., and the Pentagon. And if opposition is not enough, we must resist. And if resistance is not enough, then subvert."

Hardly had he finished speaking when Park Service police and Coconino County sheriff's deputies arrived on the scene. While they questioned Howie Wolke and me, and tried to disperse the illegal assembly, outlaw country singer Johnny Sagebrush led the demonstrators in song for another twenty minutes.

The Glen Canyon Dam caper brought Earth First! an unexpected amount of media attention. Membership quickly spiraled to more than 1,000, with members from Maine to Hawaii. Even the government became interested. According to reports from friendly park rangers, the FBI dusted the entire Glen Canyon Dam crack for fingerprints!

When a few of us kicked off Earth First!, we sensed a growing environmental radicalism in the country, but we did not expect the response we received. Maybe Earth First! is in the right place at the right time.

The cynical may smirk. "But what can you really accomplish? How can you fight Exxon, Coors, the World Bank, Japan, and the other great corporate giants of the Earth? How, indeed, can you fight the dominant dogmas of Western civilization?"

Perhaps it *is* a hopeless quest. But one who loves Earth can do no less. Maybe a species will be saved or a forest will go uncut or a dam will be torn down. Maybe not. A monkeywrench thrown into the gears of the machine may not stop it. But it might delay it, make it cost more. And it feels good to put it there.

Boundless Bull

Herman E. Daly*

If you want to know what is wrong with the American economy it is not enough to go to graduate school, read books, and study statistical trends—you also have to watch TV. Not the Sunday morning talking-head shows or even documentaries, and especially not the network news, but the really serious stuff—the commercials. For instance, the most penetrating insight into the American economy by far is contained in the image of the bull that trots unimpeded through countless Merrill Lynch commercials.

One such ad opens with a bull trotting along a beach. He is a very powerful animal—nothing is likely to stop him. And since the beach is empty as far as the eye can see, there is nothing that could even slow him down. A chorus in the background intones: "to . . . know . . . no . . . boundaries. . . ." The bull trots off into the sunset.

Abruptly the scene shifts. The bull is now trotting across a bridge that spans a deep gorge. There are no bicycles, cars or eighteen-wheel trucks on the bridge, so again the bull is alone in an empty and unobstructed world. The chasm, which might have proved a barrier

Gannett Center Journal, Summer 1990.

* The views presented here are those of the author and should in no way be attributed to the World Bank.

to the bull, who after all is not a mountain goat, is conveniently spanned by an empty bridge.

Next the bull finds himself in a forest of giant redwoods, looking just a bit lost as he tramples the underbrush. The camera zooms up the trunk of a giant redwood whose top disappears into the shimmering sun. The chorus chirps on about a "world with no boundaries."

Finally we see the bull silhouetted against a burgundy sunset, standing in solitary majesty atop a mesa overlooking a great empty southwestern desert. The silhouette clearly outlines the animal's genitalia, making it obvious even to city slickers that this is a bull, not a cow. Fadeout. The bull cult of ancient Crete and the Indus Valley, in which the bull god symbolized the virile principle of generation and invincible force, is alive and well on Wall Street.

The message is clear: Merrill Lynch wants to put you into an individualistic, macho, world without limits—the U.S. economy. The bull, of course, also symbolizes rising stock prices and unlimited optimism, which is ultimately based on this vision of an empty world where strong, solitary individuals have free reign. This vision is what is most fundamentally wrong with the American economy. In addition to TV commercials it can be found in politicians' speeches, in economic textbooks, and between the ears of most economists and business journalists.

No bigger lie can be imagined. The world is not empty; it is full! Even where it is empty of people it is full of other things. In California it is so full that people shoot each other because freeway space is scarce. A few years ago they were shooting each other because gasoline was scarce. Reducing the gasoline shortage just aggravated the space shortage on the freeways.

Many species are driven to extinction each year due to takeover of their "empty" habitat. Indigenous peoples are relocated to make way for dams and highways through "empty" jungles. The "empty" atmosphere is dangerously full of carbon dioxide and pollutants that fall as acid rain.

Unlike Merrill Lynch's bull, most do not trot freely along empty beaches. Most are castrated and live their short lives as steers imprisoned in crowded, stinking feed lots. Like the steers, we too live in a world of imploding fullness. The bonds of community, both moral and biophysical, are stretched, or rather compressed, to the breaking point. We have a massive foreign trade deficit, a domestic federal deficit, unemployment, declining real wages, and inflation. Large accumulated debts, both foreign and domestic, are being used to finance consumption, not investment. Foreign ownership of the U.S. economy is increasing, and soon domestic control over national economic life will decrease.

Why does Merrill Lynch (and the media and academia and the politicians) regale us with this "boundless bull"? Do they believe it? Why do they want you to believe it, or at least to be influenced by it at a subconscious level? Because what they are selling is growth, and growth requires empty space to grow into. Solitary bulls don't have to share the world with other creatures, and neither do you! Growth means that what you get from your bullish investments does not come at anyone else's expense. In a world with no boundaries the poor can get richer while the rich get richer even faster. Our politicians find the boundless bull cult irresistible.

The boundless bull of unlimited growth appears in economics textbooks with less colorful imagery but greater precision. Economists abstract from natural resources because they do not consider them scarce, or because they think that they can be perfectly substituted by man-made capital. The natural world either puts no obstacles in the bull's path or, if an obstacle like the chasm appears, capital (the bridge) effectively removes it.

Economics textbooks also assume that wants are unlimited. Merrill Lynch's boundless bull is always on the move. What if, like Ferdinand, he were to just sit, smell the flowers, and be content with the world as it is without trampling it underfoot? That would not do. If you are selling continual growth then you have to sell continual, restless, trotting dissatisfaction with the world as it is, as well as the notion that it has no boundaries.

This pre-analytic vision colors the analysis even of good economists, and many people never get beyond the boundless bull scenario. Certainly the media have not. Would it be asking too much of the media to do what professional economists have failed to do?

Probably so, but all disciplines badly need external critics, and in the universities disciplines do not criticize each other. Even philosophy, which historically was the critic of the separate disciplines, has abdicated that role. Who is left? Economist Joan Robinson put it well many years ago when she noted that economists have run off to hide in thickets of algebra and left the really serious problems of economic policy to be handled by journalists. Is it to the media that we must turn for disciplinary criticism, for new analytic thinking about the economy? The thought does not inspire confidence. But in the land of the blind the one-eyed man is king. If journalists are to criticize the disciplinary orthodoxy of economic growth, they will need both the energy provided by moral outrage and the clarity of thought provided by some basic analytic distinctions.

Moral outrage should result from the dawning realization that we are destroying the capacity of the Earth to support life and counting it as progress, or at best as the inevitable cost of progress. "Progress" evidently means converting as much as possible of Creation into ourselves and our furniture. "Ourselves" means, concretely, the unjust combination of overpopulated slums and overconsuming suburbs. Since we do not have the courage to face up to sharing and population control as the solution to injustice, we pretend that further growth will make the poor better off instead of simply making the rich richer. The wholesale extinctions of other species, and some primitive cultures within our own species, are not reckoned as costs. The intrinsic value of other species, their own capacity to enjoy life, is not admitted at all in economics, and their instrumental value as providers of ecological life-support services to humans is only dimly perceived. Costs and benefits to future humans are routinely discounted at 10 percent, meaning that each dollar of cost or benefit fifty years in the future is valued at less than a penny today.

But just getting angry is not sufficient. Doing something requires clear thinking, and clear thinking requires calling different things by different names. The most important analytic distinction comes straight from the dictionary definitions of growth and development. "To grow" means to increase in size by the accretion or assimilation of material. "Growth" therefore means a quantitative increase in the scale of the physical dimensions of the economy. "To develop" means to expand or realize the potentialities of; to bring gradually to a fuller, greater, or better state. "Development" therefore means the qualitative improvement in the structure, design, and composition of the physical stocks of wealth that results from greater knowledge, both of technique and of purpose. A growing economy is getting bigger; a developing economy is getting better. An economy can therefore develop without growing, or grow without developing. A steady-state economy is one that does not grow, but is free to develop. It is not static—births replace deaths and production replaces depreciation, so that stocks of wealth and people are continually renewed and even improved, although neither is growing. Consider a steady-state library. Its stock of books is constant but not static. As a book becomes worn out or obsolete it is replaced by a new or better one. The quality of the library improves, but its physical stock of books does not grow. The library develops without growing. Likewise the economy's physical stock of people and artifacts can develop without growing.

The advantage of defining growth in terms of change in physical scale of the economy is that it forces us to think about the effects of a change in scale and directs attention to the concept of an ecologically sustainable scale, or perhaps even of an optimal scale. The scale of the economy is the product of population times per capita resource use—i.e., the total flow of resources—a flow that might conceivably be ecologically unsustainable, especially in a finite world that is not empty.

The notion of an optimal scale for an activity is the very heart of microeconomics. For every activity, be it eating ice cream or making shoes, there is a cost function and a benefit function, and the rule is to increase the scale of the activity up to the point where rising marginal cost equals falling marginal benefit—i.e., to where the desire for another ice cream is equal to the desire to keep the money for something else, or the extra cost of making another pair of shoes is just equal to the extra revenue from selling the shoes. Yet for the macro level, the aggregate of all microeconomic activities (shoe making, ice cream eating, and everything else), there is no concept of an optimal scale. The notion that the macro economy could become too large relative to the ecosystem is simply absent from macroeconomic theory. The macro economy is supposed to grow forever. Since GNP adds costs and benefits together instead of comparing them at the margin, we have no macro level accounting by which an optimal scale could be identified. Beyond a certain scale growth begins to destroy more values than it creates—economic growth gives way to an era of anti-economic growth. But GNP keeps rising, giving us no clue as to whether we have passed that critical point!

The apt image for the U.S. economy, then, is not the boundless bull on the empty beach, but the proverbial bull in the china shop. The boundless bull is too big and clumsy relative to its delicate environment. Why must it keep growing when it is already destroying more than its extra mass is worth?

Because: (1) We fail to distinguish growth from development, and we classify all scale expansion as "economic growth" without even recognizing the possibility of "anti-economic growth"—i.e., growth that costs us more than it is worth at the margin; (2) we refuse to fight poverty by redistribution and sharing, or by controlling our own numbers, leaving "economic" growth as the only acceptable cure for poverty. But once we are beyond the optimal scale and growth makes us poorer rather than richer, even that reason becomes absurd. Sharing, population control, and true qualitative development are difficult. They are also collective virtues that for the most part cannot be attained by individual action and that do not easily give rise to increased opportunities for private profit. The boundless bull is much easier to sell, and profitable at least to some while the illusion lasts. But further growth has become destructive of community, the environment, and the common good. If the media could help economists and politicians to see that, or at least to entertain the possibility that such a thing might be true, they will have rendered a service far greater than all the reporting of statistics on GNP growth, Dow Jones indexes, and junk bond prices from now until the end of time.

Grass-Roots Groups Are Our Best Hope for Global Prosperity and Ecology

Alan B. Durning

Women on the banks of the Ganges may not be able to calculate an infant mortality rate, but they know all too well the helplessness and agony of holding a child as it dies of diarrhea. Residents along the lower reaches of the Mississippi may not be able to name the carcinogens and mutagens that nearby chemical factories pump into the air and water, but they know how many of their neighbors have suffered miscarriages or died of cancer. Forest dwellers in the Amazon basin cannot quantify the mass extinction of species now occurring around them, but they know what it is to watch their primeval homeland go up in smoke before advancing waves of cattle ranches and developers.

These men and women in Bangladesh, Louisiana, and Brazil understand environmental degradation in its rawest forms. To them, creeping destruction of ecosystems has meant deteriorating health, lengthening workdays, and failing livelihoods. And it has pushed

Utne Reader, July/August 1989.

many of them to act. In villages, neighborhoods, and shantytowns around the world, people are coming together to challenge the forces of environmental and economic decline that endanger our communities and our planet.

In the face of such enormous threats, isolated grass-roots organizing efforts appear minuscule—ten women plant trees on a roadside, a local union strikes for a non-toxic workplace, an old man teaches neighborhood children to read—but, when added together, their impact has the potential to reshape the earth. Indeed, local activists form a front line in the worldwide struggle to end poverty and environmental destruction. These grass-roots groups include workplace co-ops, suburban parents committees, peasant farmers unions, religious study groups, neighborhood action federations, tribal nations, and innumerable others. Although widely diverse in origins, these groups share a common capacity to utilize local knowledge and resources, to respond to problems rapidly and creatively, and to maintain the flexibility necessary to adapt to changing circumstances. In addition, although few groups use the term sustainable development, their agendas often embody this ideal. They want economic prosperity without sacrificing their health or the prospects for their children.

All by itself this new wave of local activist organizations is nowhere near powerful enough to shift modern industrial society onto a more sustainable course of development. The work required—from slowing excessive population growth to reforesting the planet—will involve an unprecedented outpouring of human energy. Yet community groups, whose membership now numbers in the hundreds of millions worldwide, may be able to show the world how to tap the energy to perform these acts.

Grass-roots action is on the rise everywhere, from Eastern Europe's industrial heartland, where fledgling environmental movements are demanding that human health no longer be sacrificed for economic growth, to the Himalayan foothills, where multitudes of Indian villagers are organized to protect and reforest barren slopes. As environmental decay accelerates in industrial regions, local communities are organizing in growing numbers to protect themselves from chemical wastes, industrial pollution, and nuclear power installations. Meanwhile, in developing countries, deepening poverty combined with often catastrophic ecological degradation has led to the proliferation of grass-roots self-help movements.

In the Third World, especially, traditional tribal, village, and religious organizations—first disturbed by European colonialism—have been stretched and often dismantled by the great cultural upheavals of the twentieth century: rapid population growth, urbanization, the advent of modern technology, and the spread of Western commercialism. Community groups have been formed in many places to meet the economic and social needs these traditional ties once fulfilled.

In the face of seemingly insurmountable problems, community groups around the planet have been able to accomplish phenomenal things.

- In Lima's Villa El Salvador district, Peruvians have planted a half-million trees; built 26 schools, 150 daycare centers, and 300 community kitchens; and trained hundreds of door-to-door health workers. Despite the extreme poverty of the district's inhabitants and a population that has shot up to 300,000, illiteracy has fallen to 3 percent—one of the lowest rates in Latin America—and infant mortality is 40 percent below the national average. The ingredients of success have been a vast network of women's groups and the neighborhood association's democratic administrative structure, which extends down to representatives on each block.
- In Dhandhuka, on the barren coastal plain of India's Gujarat state, a generation of excessive fuelwood gathering and overgrazing has led to desertification, which in turn has triggered social and economic disintegration. As cattle died of thirst, the children lost their milk, making them easy victims for the diseases that prey on the malnourished. Conflicts erupted over water that seeped into brackish wells, and in the worst years, four-fifths of the population had to migrate to survive. As in much of the world, fetching water in Dhandhuka is women's work. Thus it was the women who decided, upon talking with community organizers in 1981, to build a permanent reservoir to trap the seasonal rains. Migrant laborers described irrigation channels lined with plastic sheets they had seen elsewhere, and the villagers reasoned that a reservoir could be sealed the same way.

> After lengthy discussion and debate, the community agreed to the plan, and in 1986, all but a few stayed home during the dry season to get the job done. Moving thousands of tons of earth by hand, they finished the pool before the rains returned. The next dry season they were well-supplied, which inspired neighboring villages to plan their own reservoirs.

Asia has by many accounts the most active grass-roots movement. India's self-help movement has a prized place in society, tracing its roots to Mahatma Gandhi's pioneering village development work sixty years ago. Gandhi aimed to build a just and humane society from the bottom up, starting with self-reliant villages based on renewable resources. Tens if not hundreds of thousands of local groups in India now wage the day-by-day struggle for development.

Across the region, community activism runs high. Three million Sri Lankans, for instance, participate in Sarvodaya Shramadana, a community development movement that combines Gandhian teachings with social action tenets of Theravada Buddhism. Sarvodaya mobilizes massive work teams to do everything from building roads to draining malarial ponds.

After Asia, Latin American communities are perhaps the most active. The event that sparked much of this work was the 1968 conference of Catholic Bishops in Medellín, Colombia, where the Roman Catholic Church fundamentally reoriented its social mission toward improving the lot of the poor. Since that time, millions of priests, nuns, and laypeople have fanned out into the back streets and hinterlands from Tierra del Fuego to the Rio Grande, dedicating themselves to creating a people's church embodied in neighborhood worship and action groups called Christian Base Communities. Brazil alone has 100,000 base communities, with at least 3 million members, and an equal number are spread across the rest of the continent.

Latin American political movements also laid the groundwork for current community self-help efforts. A decade ago, the rise and subsequent repression of Colombia's National Association of Small Farmers gave peasants experience with organizing that led to the abundance of community efforts today, including cooperative stores and environmental "green councils." In Nicaragua, the national uprising that overthrew the dictatorship of Anastasio Somoza in 1979 created a surge of grass-roots energy that flowed into thousands of new cooperatives, women's groups, and community-development projects.

Self-help organizations are relative newcomers to Africa, though traditional village institutions remain stronger here than in other regions. Nevertheless, in parts of Africa where political struggles have led to dramatic changes in political structures, local initiatives have sprung up in abundance. In Kenya, the *harambee* (let's pull together) movement began with independence in 1963 and, with encouragement from the national government, by the early '80s was contributing nearly one-third of all labor, materials, and finances invested in rural development. With Zimbabwe's transfer to black rule in 1980, a similar explosion in community organizing began, as thousands of women's community gardens and informal small-farmer associations formed.

A noteworthy characteristic of community movements throughout the Third World is the central role that women play. Women's traditional nurturing role may give them increased concern for the generations of their children and grandchildren, while their subordinate social status gives them more to gain from organizing.

Unfortunately, the map of Third World local action has several blank spaces. Independent community-level organizations concentrating on self-help are scarce or non-existent in the Middle East, China, north Africa, large patches of sub-Saharan Africa, and northeastern India. Likewise, remote regions in many countries lack grass-roots groups. Some of these absences are a result of cultural, religious, or political factors, as in China, where state-sanctioned local groups monopolize grass-roots development. Northeastern India and sub-Saharan Africa, by contrast, are home to some of the poorest people on earth. The absence of local groups there may reflect a degree of misery that prevents energy from being expended on anything beyond survival.

Outside the Third World, grass-roots movements are also on the rise. In the Soviet Union and Eastern Europe, where officially sanctioned local organizations are numerous but largely controlled by state and party hierarchies, the political openness of this decade has brought a wave of independent citizens groups. (See "Independent activists challenge the status quo in Eastern Europe," *Utne Reader*, Jan./Feb. 1989.)

Environmental issues and grass-roots politics play a major role in the new nationalist movements rocking the USSR. In February 1988, thousands of Armenians, tired of bearing the brunt of pollution from the scores of local chemical facilities, demanded cancellation of a proposed plant near their capital city, Yerevan. Eight months later, 50,000 Latvians, Estonians, and Lithuanians linked arms in a human chain stretching 150 kilometers along the shore of the severely polluted Baltic Sea to protest Soviet planners' blatant disregard for the ecology of their homelands.

In those regions where nuclear power is still on a growth course—Japan, France, and Eastern Europe—anti-nuclear movements have grown dramatically since the 1986 explosion at Chernobyl. Intense popular opposition seems to follow nuclear power wherever it goes. In the Soviet Union, public protests have led to plans to close one operating nuclear reactor and to cancellation of at least five planned plants. In Japan, an unprecedented groundswell—the first nationwide movement on an environmental issue in the country's history—has enrolled tens of thousands of citizens with no past experience in political activism. Women in particular are joining in large numbers, apparently sensitized by fears of radioactive food imported from Europe after Chernobyl.

In Western industrial nations, community-based organizations set their sights on everything from local waste recycling to international trade and debt issues. The ascent of the German Green Party in the early '80s was partly a product of an evolution in this movement from local to national concerns. Inspired by their German counterpart, Green parties have sprung up in sixteen European countries and already hold parliamentary seats in more than half of them.

Paralleling a steady rise in neighborhood organizing on local social and economic issues, the U.S. environmental movement experienced a marked grassroots expansion in the early '80s. Local concern focuses particularly on toxic waste management, groundwater protection, and solid waste problems.

Issue-oriented environmental activism is not peculiar to industrial lands. Just as grass-roots self-help movements have spread through the slums and countrysides of many developing nations, vocal advocates for environmental protection have emerged in most capital cities. Malaysia, India, Brazil, Argentina, Kenya, Mexico, Indonesia, Ecuador, Thailand, and other developing countries have all given birth to activist groups—largely since 1980. Sri Lanka alone has a congress of environmental groups with 100 members. Environmental movements and grass-roots development movements have also begun to work together.

Despite the heartening rise of grass-roots action, humanity is losing the struggle for sustainable development. For every peasant movement that reverses the topsoil erosion of valuable agricultural land, dozens more fail. For each neighborhood that rallies to replace a proposed garbage burner with a recycling program, many others remain mired in inaction. Spreading today's grass-roots mobilization to a larger share of the world's communities is a crucial step toward putting an end to the global scourges of poverty and environmental degradation.

All local groups eventually collide with forces they cannot control. Peasant associations cannot enact national or international agricultural policies or build roads to distant markets. Women's groups cannot develop and test modern contraceptive technologies or rewrite bank lending rules. Neighborhood committees cannot implement citywide recycling programs or give themselves a seat at the table in national energy planning. The greatest obstacle to community action is that communities cannot do it alone. Small may be beautiful, but it can also be insignificant.

The prospects for grass-roots progress against poverty are further limited in a world economy in which vested interests are deeply entrenched and power is concentrated in a few nations. Thus reforms at the international level are as important as those in the village.

The largest challenge in reversing global ecological and economic deterioration is to forge an alliance between local groups and national governments. Only governments have the resources and authority to create the conditions for full-scale grass-roots mobilization. In the rare cases where national-local alliances have been forged, extraordinary gains have followed. South Korea and China have used village-level organizations to plant enormous expanses of trees, implement national population policies, and boost agricultural production. Zimbabwe has trained more than 500 community-selected family planners to improve maternal and child health and control population growth. In the year after the 1979 Nicaraguan revolution, a massive literacy campaign sent 90,000 volunteers into the countryside; in one year, they raised literacy from 50 to 87 percent. Even under Ferdinand Marcos' repressive rule in the Philippines, the National Irrigation Administration amazingly transformed itself into a people-centered institution, cooperating with peasant associations.

Full-scale community-state alliances can come about only when a motivated and organized populace joins forces with responsive national leadership. But herein lies the greatest obstacle to mobilizing for prosperity and ecology: Few leaders are committed to promoting popular organizations. Because government's first concern is almost always to retain political power, independent-minded grass-roots movements are generally seen as more of a threat than an ally. Unrepresentative elites rule many nations and all too often they crush popular movements rather than yield any of their privilege or power. Inevitably, self-help movements will clash with these forces, because like all development, self-help is inherently political: It is the struggle to control the future.

Essentially, grass-roots action on poverty and the environment comes down to the basic question of people's right to shape their own destiny. Around the world, community organizations are doing their best to put this participatory vision into practice, and they are simultaneously posing an even deeper question. In the world's impoverished south, it is phrased, "What is development?" In the industrial north, "What is progress?" Behind the words, however, is the same profoundly democratic refrain: What kind of society shall our nation be? What kind of lives shall our people lead? What kind of world shall we leave to our children?

Whether these scattered beginnings launched by grass-roots groups eventually rise in a global groundswell depends only on how many more individuals commit their creativity and energy to the challenge. The inescapable lesson for each of us is distilled in the words of Angeles Serrano, a grandmother and community activist from the slums of Manila. "Act, act, act. You can't just watch."

The Consummate Recycler

Larry Borowsky

Cliff Humphrey took a sledge hammer to his car in 1969. The public dismantling of the 1958 pollution-spewing Rambler (the last car Cliff and his wife Mary have owned) showed the depth of Cliff's commitment to a healthier environment.

Other people wouldn't go so far to protect the earth, Humphrey knew, but there was a less traumatic option—recycling.

Humphrey recognized that the small chore of separating recyclables from out-going trash could give citizens a sense of personal responsibility for the quality of the environment. Thus inspired, people might eliminate all kinds of wasteful habits.

Twenty years ago, few cities paid attention when Cliff Humphrey suggested that newspapers, bottles, and cans could be systematically collected for recycling. These days, communities across the United States have made recycling a high priority, and many are hiring Humphrey's services to help launch local recycling programs.

This irony provides Humphrey, now aged 51, with no small measure of amusement, but he's not the "I-told-you-so" type.

In these two decades of surge-and-slough environmental concern,

Environmental Action, March/April 1989.

Humphrey has been one of recycling's most vital and dedicated practitioners.

Humphrey's trend-setting grassroots organization, Ecology Action, set up one of the country's first recycling programs in Berkeley, California. Soon after, Humphrey introduced a pioneering concept—curbside pickup of recyclable trash—to a nearby city. In a salute to his accomplishments and longevity, Humphrey was introduced as the "grandfather of recycling" at the 1988 National Recycling Congress, an annual meeting of recycling activists and administrators.

"He's always been a hero of mine," says Neil Seldman, who is himself a recycling leader.

"He's taught me a lot. The reason Cliff stands out [from other recyclers] is because they are technicians. Cliff has the technical know-how, but he also has a philosophy," says Seldman, director of the Institute for Local Self-Reliance in Washington, D.C.

"Cliff is an innovator," adds San Diego County recycling coordinator Rick Anthony.

In 1968, the Humphreys and another couple founded Ecology Action. An art exhibition on the world's environmental ills—created by artist Mary—brought in the first members.

It was a loose coalition of individuals seeking to teach the public how to conserve resources and avoid polluting the environment. Members carried groceries home in knapsacks, rather than the standard supermarket brown bag. They avoided excessive throwaway packaging, and piled leftover food atop a compost heap instead of into the trash can. Old clothes were patched up, envelopes used and reused and re-reused. Cliff and Mary ceremoniously dismembered their car, and then members took a 500-mile "survival walk" through California for Earth Day 1970.

Cliff, who had been a highway engineer in the army and studied ecology at the University of California, began to push recycling.

For guidance, he looked to the emergency recycling efforts of World War II and the commercial "salvage" industry. Profit in the salvage business depended on pursuing the large, clean, easy-to-collect lumps of waste churned out by commercial establishments.

Since Ecology Action wasn't in it for the money, but to change individual attitudes toward waste, the group resolved to pursue the more undisciplined garbage produced by ordinary people.

This would be far more problematic. Commercial waste could be collected efficiently because it was generated at relatively few sites. Residential waste was generated in small amounts at every house in town. If it solved this problem, Ecology Action would still have to find markets for the recovered materials. And, of course, the proposition couldn't work at all unless residents went to the trouble of keeping recyclable items out of their trash.

"Nobody back then believed that people would fuss with their garbage for environmental reasons," Humphrey recalls. But he had faith that the public would cooperate.

And people did. Early in 1970, Ecology Action began collecting newspapers, bottles, and cans every weekend in fifty-five gallon drums set up in a parking lot. Almost immediately, Berkeley's "drop-off" recycling center touched off a craze. One observer likened the weekly scene to an exuberant county fair. Hippies, little old ladies, society matrons, and businessmen stood shoulder to shoulder with their boxes and bags of refuse while Ecology Action volunteers handled the drums.

Tremendous volumes of material flowed in every week, swallowing up the group's storage capacity. The group hustled to find buyers for the materials and were pleasantly surprised by the cooperative spirit some local industries displayed. One company even loaned Ecology Action a truck and driver once a week.

Citizens in neighboring communities soon began to organize recycling programs of their own; one year later a San Francisco newspaper counted seventy-one ongoing recycling efforts in the Bay Area.

The Berkeley program won many admirers, but it had its detractors as well. The critics asserted that dropoff recycling produced negligible environmental benefits. Humphrey recalls one study, conducted by Stanford University engineers, which concluded that the amount of gasoline wasted (and air pollution produced) by people making special trips to drop off their recyclables outweighed the energy and resources saved by recycling.

Humphrey freely concedes that dropoff centers aren't the most environmentally beneficial approach to recycling. "The dropoff was a necessary first, inefficient step," he explains. "It was the way that you established the recycling values in the community."

And changing public values was the #1 priority.

Rapid success with Ecology Action's dropoff experiment did not fully satisfy Humphrey. He realized that Berkeley's population, a funky mix of intellectuals and hippies, had an unusual predisposition to try out new ideas. Recycling would need a broader constituency. Could it work in a more typical American city?

In the summer of 1970, the Humphreys withdrew from the booming Berkeley recycling scene and moved inland to Modesto, California, a conservative agricultural town of about 50,000 residents. Arriving with no money, Cliff and Mary lived temporarily in a tent pitched in a friend's backyard. With a few loans, donations, and receipts from the sales of Cliff's elementary school textbook, *What's Ecology?* they soon launched a new chapter of Ecology Action.

Humphrey took his case directly to the public, telling the local media that he intended to turn Modesto into a "model of ecological sanity" for the rest of the nation. Members of the community hardly rejoiced over this news.

"You had a lot of suspicion that a long-haired Berkeley radical was importing something weird from the Bay Area," recalls Peggy Mensinger, an early Humphrey supporter who served as the city's mayor from 1979 to 1987.

But Humphrey's high profile in the local papers and on the radio gradually began to convince people that recycling was not a subversive plot, but a common sense approach to managing natural resources.

Within a year *Look* magazine reported that Modesto had gone "bananas" over recycling. The Boy Scouts, the Modesto YMCA, several church congregations, and various other groups helped round up money, volunteers, and material for Ecology Action's new dropoff recycling center. Even gas stations and bars began saving their beverage containers. The zeal with which Modesto took to recycling surprised Humphrey himself.

"There's no holding this thing back," he told *Look*. "If we make recycling work here, it will work anywhere."

Modesto's enthusiasm convinced Humphrey that the town was ready for the next phase of recycling—curbside pickup. This idea, first conceived back in the Berkeley days, would make recycling more efficient and more convenient for residents. Initially, it sparked a new flurry of negative commentary.

"Everybody told me that it just wouldn't work," he remembers. "They said if you put stuff on the street, the wind will blow the papers all over, and the kids will pick up the glass bottles and smash them."

As with all his pioneering steps, Humphrey knew he was taking a risk. "To be honest, we didn't know whether the kids were going to smash the bottles or not," he admits today. "But we went ahead."

Ecology Action put the first pilot route into operation in early 1972 with a pair of secondhand trucks while city officials fidgeted. "I think they figured, 'Well, let him do that for two years or so and then he'll go away, and our public works people can go back to putting the trash where it belongs,'" Humphrey laughs. "But people kept bringing it out to the curb and we kept picking up."

By the end of the decade, Ecology Action provided curbside service to all of Modesto and the entire surrounding region. It sent out five collection trucks every day, operated a buy-back center, and maintained dropoff sites throughout the area. Recycling had become a Modesto institution, a part of the community's daily routine.

More importantly, it had changed the outlook of the city's residents. "Cliff brought Modesto new perspectives," says Peggy Mensinger. "He created a consciousness here, a force for environmental action."

Humphrey was named "Recycler of the Decade" at the first National Recycling Congress in 1980. Ironically enough, Humphrey was struggling to keep Ecology Action afloat at the time. The non-profit group survived almost entirely on the resale of the materials it collected, so when market prices dropped sharply in the late 1970s, debts quickly began to accrue.

"At some point everyone rises to their level of incompetence," Humphrey says wryly. "This is where I discovered mine."

Ecology Action pulled through, but the wearying pace burned

Humphrey out. Turning the organization's reins over to a successor, he and Mary headed back to the Bay Area in the fall of 1980. No sooner had he "retired" from recycling, but a new opportunity came along. San Francisco city supervisors were looking for someone to put together a comprehensive municipal recycling program. The offer was too good to pass up.

But Humphrey's vision of citywide curbside pickup soon ran into an indomitable city machine. "You have to work six or seven months and finally people start showing you where the doors are to the closets," he comments ruefully.

Humphrey eventually secured state money to get the program off the ground, but he didn't stay on the job much longer. He found himself at odds with city planners who wanted to combine recycling with garbage incineration. Humphrey said no thanks.

Today there's a booming demand for people who know how to make recycling work. A critical, nationwide shortage of landfill space and skyrocketing garbage disposal costs have forced communities to reduce the volume of waste that goes to the dump. Currently employed by a private consulting firm in San Francisco, Humphrey spends weeks at a time on the road helping local officials plan and execute recycling projects.

Recycling is only beginning to hit its stride, Humphrey feels. "Right now we recover around 12 percent of the waste stream," he says. "In theory we could do 50 percent. If you compost the organic stream—leaves, yard clippings, and so forth—and recycle the cardboard, paper, and glass and metal containers, you've taken care of almost half of the overall weight."

Local and state governments are investing millions of dollars to develop recycling, which they now regard as a potential savior. Humphrey sees this as a crucial test.

"A lot is being asked of recycling right now," Humphrey cautions. Achieving recycling's potential will require more efficient collection methods and new markets for the resale of the materials collected.

It will also take a massive education campaign to make the public more diligent in its recycling habits. If Humphrey's career has proven anything, however, it is that these obstacles can be overcome.

The programs he built in Modesto and Berkeley, and helped launch in San Francisco, have become lasting successes. Half Berkeley's population recycles at least some garbage today, and the local government has set 50 percent of the waste stream as an official recycling goal to be reached by 1991. The San Francisco program, now in its ninth year, recycles about 240,000 tons of garbage annually, about 25 percent of the city's solid waste output. Modesto's recycling program was taken over by the city in 1982 and later by private contractors.

These developments meet with Humphrey's approval. But he is most pleased by the overall level of public concern over environmental issues. "The newspapers wouldn't give so much space to acid rain and the greenhouse effect if they couldn't take an interested readership for granted," Humphrey says. "I'd like to think that all the years of recycling have helped bring that along."

As in conservative Modesto, recycling has gotten the country onto the right path, Humphrey feels, but there is still a lot of ground to cover. He points out that America still needs to burn less gasoline, use less electricity, and stop dumping so many toxics into air and water.

If it has taken twenty years and a solid waste "crisis" for such a painless adjustment as recycling to gain acceptance, will Americans make bigger sacrifices for environmental well-being?

"I know that these ideas can catch on," Humphrey insists. "I've seen it happen with recycling. It will take time. But it can be done."

From The Dream of the Earth

Thomas Berry

Returning to Our Native Place

We are returning to our native place after a long absence, meeting once again with our kin in the earth community. For too long we have been away somewhere, entranced with our industrial world of wires and wheels, concrete and steel, and our unending highways, where we race back and forth in continual frenzy.

The world of life, of spontaneïty, the world of dawn and sunset and glittering stars in the dark night heavens, the world of wind and rain, of meadow flowers and flowing streams, of hickory and oak and maple and spruce and pineland forests, the world of desert sand and prairie grasses, and within all this the eagle and the hawk, the mockingbird and the chickadee, the deer and the wolf and the bear, the coyote, the raccoon, the whale and the seal, and the salmon

The Dream of the Earth (San Francisco: Sierra Club Books, 1988).

returning upstream to spawn—all this, the wilderness world recently rediscovered with heightened emotional sensitivity, is an experience not far from that of Dante meeting Beatrice at the end of the *Purgatorio*, where she descends amid a cloud of blossoms. It was a long wait for Dante, so aware of his infidelities, yet struck anew and inwardly "pierced," as when, hardly out of his childhood, he had first seen Beatrice. The "ancient flame" was lit again in the depths of his being. In that meeting, Dante is describing not only a personal experience, but the experience of the entire human community at the moment of reconciliation with the divine after the long period of alienation and human wandering away from the true center.

Something of this feeling of intimacy we now experience as we recover our presence within the earth community. This is something more than working out a viable economy, something more than ecology, more even than Deep Ecology, is able to express. This is a sense of presence, a realization that the earth community is a wilderness community that will not be bargained with; nor will it simply be studied or examined or made an object of any kind; nor will it be domesticated or trivialized as a setting for vacation indulgence, except under duress and by oppressions which it cannot escape. When this does take place in an abusive way, a vengeance awaits the human, for when the other living species are violated so extensively, the human itself is imperiled.

If the earth does grow inhospitable toward human presence, it is primarily because we have lost our sense of courtesy toward the earth and its inhabitants, our sense of gratitude, our willingness to recognize the sacred character of habitat, our capacity for the awesome, for the numinous quality of every earthly reality. We have even forgotten our primordial capacity for language at the elementary level of song and dance, wherein we share our existence with the animals and with all natural phenomena. Witness how the Pueblo Indians of the Rio Grande enter into the eagle dance, the buffalo dance, and the deer dance; how the Navajo become intimate with the larger community through their dry-paintings and their chantway ceremonies; how the peoples of the Northwest express their identity through their totem animals; how the Hopi enter into communication with desert rattlesnakes in their ritual dances. This mutual presence finds expression also in poetry and in story form, especially in the trickster stories of the Plains Indians in which Coyote performs his never-ending magic. Such modes of presence to the living world we still carry deep within ourselves, beyond all the suppressions and even the antagonism imposed by our cultural traditions.

Even within our own Western traditions at our greater moments of expression, we find this presence, as in Hildegard of Bingen, Francis of Assisi, and even in the diurnal and seasonal liturgies. The dawn and evening liturgies, especially, give expression to the natural phenomena in their numinous qualities. Also, in the bestiaries of the medieval period, we find a special mode of drawing the animal world into the world of human converse. In their symbolisms and especially in the moral qualities associated with the various animals, we find a mutual revelatory experience. These animal stories have a playfulness about them, something of a common language, a capacity to care for each other. Yet these movements toward intensive sharing with the natural world were constantly turned aside by a spiritual aversion, even by a sense that humans were inherently cut off from any true sharing of life. At best they were drawn into a human context in some subservient way, often in a derogatory way, as when we projected our own vicious qualities onto such animals as the wolf, the rat, the snake, the worm, and the insects. We seldom entered their wilderness world with true empathy.

The change has begun, however, in every phase of human activity, in all our professions and institutions. Greenpeace on the sea and Earth First! on the land are asserting our primary loyalties to the community of earth. The poetry of Gary Snyder communicates something of the "wild sacred" quality of the earth. In his music Paul Winter is responding to the cry of the wolf and the song of the whale. Roger Tory Peterson has brought us intimately into the world of the birds. Joy Adamson has entered into the world of the lions of Africa; Dian Fossey the social world of the gentle gorilla. John Lilly has been profoundly absorbed into the consciousness of the dolphin. Farley Mowat and Barry Lopez have come to an intimate understanding of the gray wolf of North America. Others have learned the dance language of the bees and the songs of the crickets.

What is fascinating about these intimate associations with various living forms of the earth is that we are establishing not only an acquaintance with the general life and emotions of the various species, but also an intimate rapport, even an affective relationship, with individual animals within their wilderness context. Personal names are given to individual whales. Indeed, individual wild animals are entering into history. This can be observed in the burial of Digit, the special gorilla friend of Dian Fossey's. Fossey's own death by human assault gives abundant evidence that if we are often imperiled in the wilderness context of the animals, we are also imperiled in the disturbed conditions of what we generally designate as civilized society.

Just now one of the significant historical roles of the primal people of the world is not simply to sustain their own traditions, but to call the entire civilized world back to a more authentic mode of being. Our only hope is in a renewal of those primordial experiences out of which the shaping of our more sublime human qualities could take place. While our own experiences can never again have the immediacy or the compelling quality that characterized this earlier period, we are experiencing a postcritical naiveté, a type of presence to the earth and all its inhabitants that includes, and also transcends, the scientific understanding that now is available to us from these long years of observation and reflection.

Fortunately we have in the native peoples of the North American continent what must surely be considered in the immediacy of its experience, in its emotional sensitivities, and in its modes of expressions, one of the most integral traditions of human intimacy with the earth, with the entire range of natural phenomena, and with the many living beings which constitute the life community. Even minimal contact with the native peoples of this continent is an exhilarating experience in itself, an experience that is heightened rather than diminished by the disintegrating period through which they themselves have passed. In their traditional mystique of the earth, they are emerging as one of our surest guides into a viable future.

Throughout their period of dissolution, when so many tribes have been extinguished, the surviving peoples have manifested what seems to be an indestructible psychic orientation toward the basic structure and functioning of the earth, despite all our efforts to impose on them our own aggressive attitude toward the natural world. In our postcritical naiveté we are now in a period when we become capable once again of experiencing the immediacy of life, the entrancing presence to the natural phenomena about us. It is quite interesting to realize that our scientific story of the universe is giving us a new appreciation for these earlier stories that come down to us through peoples who have continued their existence outside the constraints of our civilizations.

Presently we are returning to the primordial community of the universe, the earth, and all living beings. Each has its own voice, its role, its power over the whole. But, most important, each has its special symbolism. The excitement of life is in the numinous experience wherein we are given to each other in that larger celebration of existence in which all things attain their highest expression, for the universe, by definition, is a single gorgeous celebratory event.

About the Contributors

Edward Abbey, who died in 1989, was a writer and a defender of wilderness whose eloquent prose has been an inspiration for environmentalists. Among his better known books are *The Monkey Wrench Gang* and *The Journey Home*. His reputation as a leading philosopher of the environmental movement continues to grow.

Jose Barreiro is a Guajiro from eastern Cuba now living in New York. He is director of the Indigenous Communications Resource Center at Cornell University and editor of the Northeast Indian Quarterly. His most recent book is *View from the Shore: American Indian Perspectives on the Quincentenary*.

Thomas Berry is an historian of cultures and a past president of the American Teilhard Association for the Human Future. He is widely published. He is presently director of New York's Riverdale Center of Religious Research.

Wendell Berry is a poet, essayist, novelist, and farmer. He has taught at a number of universities including Stanford and New York University. He presently lives on a farm in Kentucky and recently published his eleventh collection of essays, *What Are People For?*

Joan Bird received her doctorate in zoology from the University of Montana. Her graduate research was on islands in the West Indies. She presently works for The Nature Conservancy in Helena, Montana.

Larry Borowsky is a freelance writer living in Denver, Colorado. His work has appeared in a number of alternative periodicals. He is presently an editor at *Westword*, an alternative weekly newspaper.

Barry Commoner has been a prominent figure in the environmental movement since the 1960s. *Time* magazine referred to him as the "Paul Revere of Ecology." He is currently the director of the Center for the Biology of Natural Systems. His most recent book is *Making Peace with the Planet*.

Herman E. Daly is senior economist, Environmental Department of the World Bank. He has been an adviser to numerous environmental organizations, and is co-founder and associate editor of the journal *Ecological Economics*. He is the author of the book *Steady-State Economics*.

Wayne H. Davis is a professor of zoology at the University of Kentucky. His research interests focus on the natural history of birds and mammals. He has published widely on a variety of biological topics including population growth and has testified before Congress on human population problems.

Alan B. Durning is a senior researcher at Worldwatch Institute in Washington, D.C. He specializes in issues of the relationship between economic inequality and environmental degradation and has written widely on the subject. In the 1989 World Hunger Media Awards he was awarded first prize for periodical coverage.

David Ehrenfeld is a professor of biology at Cook College, Rutgers University. He has written widely in the area of conservation and environmental philosophy and is author of the book *The Arrogance of Humanism*. He founded the *Journal of Conservation Biology*, which he presently edits.

Anne Ehrlich is associate director of the Center for Conservation Biology at Stanford University. She has written widely on population biology and environmental protection. She has served as a consultant to the White House Council on Environmental Quality and is presently on the board of the Center for Innovative Diplomacy.

Paul Ehrlich is Bing Professor of Population Studies at Stanford University. He has authored hundreds of articles and thirty books, among them *The Population Bomb*. Presently he is president of the American Institute of Biological Sciences, and he is also a recent recipient of a MacArthur Fellowship.

Stuart Ewen is professor of media services at Hunter College and a professor at the City of New York Graduate Center. His recent book *All-Consuming Images: The Politics of Style in Contemporary Culture* provided the basis for the award-winning PBS series *The Public Mind*.

Dave Foreman worked for the Wilderness Society during the 1970s. He helped found Earth First! in 1980 and edited the *Earth First! Journal from* 1982 to 1988. He is the author of *Ecodefense: A Field Guide to Monkeywrenching* and operates a mail-order bookstore, Dave Foreman's Conservation Bookshelf.

Chellis Glendinning is a psychologist, lecturer, and author of *Waking Up in the Nuclear Age* and *When Technology Wounds*. She serves on the council of the ecological think-tank the Elmwood Institute, is on the board of Interhelp, and is an associate of the Earth Island Institute.

James E. Lovelock is an atmospheric scientist who makes his home in Cornwall, England. He has spent years in a variety of American institutions, including both Harvard and Yale. He is the father of the Gaia Hypothesis, which he first advanced in 1972 and about which he has written widely.

Oren Lyons is a chief of the Onondaga Nation, an artist, and a professor in the American Studies Program of the State University of New York at Buffalo. He is also the publisher of *Daybreak*, a periodical concerning Native American viewpoints.

Carolyn Merchant is a professor of environmental history, philosophy, and ethics in the Department of Conservation and Resource Studies at the University of California at Berkeley. She is the author if *Ecological Revolutions: Nature, Gender, and Science in New England*, and of *The Death of Nature: Women, Ecology, and the Scientific Revolution*.

Norman Myers is a British consultant on the environment and development. He has visited Amazonia eight times over the past fifteen years and is author of *The Primary Source: Tropical Forests and Our Future*. In 1986, he was awarded the New York Zoological Society's Gold Medal for conservation.

Reed F. Noss received his Ph.D. in wildlife ecology. He is a private consultant in ecology and conservation biology. His areas of specialization are conservation planning and management at a landscape scale.

Judith Plant is an ecofeminist who is active in bioregionalism both at the local level and in the North American Bioregional Congress. She has written extensively on ecofeminist and bioregional issues and edited the anthology *Healing the Wounds: The Promise of Ecofeminism*. She co-edits *The New Catalyst*, a journal for social and political change.

Van Rensselaer Potter is a past co-director of the McArdle Laboratory, a past president of the American Society of Cell Biology, and a recipient of the American Cancer Society's Medal of Honor. He is presently emeritus professor at the University of Wisconsin. He is the author of the 1988 book *Global Bioethics*.

Paul Shepard is Avery Professor of Human Ecology at Pitzer College of Claremont College in California. Besides *Nature and Madness*, quoted in this volume, his most recent books are *The Sacred Paw: The Bear in Nature, Myth and Literature*, and a new edition of *Man in the Landscape: A Historic View of the Esthetics of Nature*.

Wallace Stegner, a writer of fiction, history, biography, and essays is presently a member of the Governing Council of The Wilderness Society. His novel *Angle of Repose* won the Pulitzer Prize in 1972, and *The Spectator Bird* won the 1977 National Book Award. His latest book is *Collected Stories*.

Edward O. Wilson is Baird Professor of Science at Harvard University. He has published widely, and among his many books are *Sociobiology: The New Synthesis* and *Biophilia*. He is a winner of a Pulitzer Prize and a National Medal of Science.

About the Editor

Bill Willers was born in Vancouver, Washington, and while growing up lived in many parts of the United States, in Europe, and in Japan. He is professor of biology at the University of Wisconsin at Oshkosh, an adviser to student environmental groups, and is involved in the Green Party movement. He is the author of *Trout Biology: A Natural History of Trout and Salmon*.

Permissions

"Arizona," by Edward Abbey, is from his book *One Life at a Time, Please.* Copyright © 1978, 1983, 1984, 1985, 1986, 1988 by Edward Abbey. Reprinted by permission of Henry Holt and Company, Inc.

"Every River I Touch Turns to Heartbreak," by Edward Abbey, appeared in a special edition of *Environmental Action* magazine titled "Visions" in 1985. Reprinted by permission of *Environmental Action.*

"Indigenous Peoples Are the 'Miner's Canary' of the Human Family," by Jose Barreiro, appeared in 1985 in a special edition of *Environmental Action* magazine titled "Visions." Reprinted by permission of *Environmental Action.*

"Returning to Our Native Place," by Thomas Berry, is from his 1988 book *The Dream of the Earth.* Reprinted by permission of Sierra Club Books.

"The Futility of Global Thinking" © 1989 by Wendell Berry is adapted from "Word and Flesh," a commencement address, and originally appeared in *Harper's* magazine, September 1989. Excerpted from *What Are People For?* by Wendell Berry, published by North Point Press, 1990.

"How Big Is Big Enough?," by Joan Bird, first appeared in the July 1989 edition of *Northern Lights* magazine. Reprinted by permission of the author and of the Northern Lights Research and Education Institute.

"The Consummate Recycler," by Larry Borowsky, appeared in the March/April 1989 edition of *Environmental Action* magazine. Reprinted by permission of *Environmental Action.*

"Why We Have Failed," by Barry Commoner, appeared in *Greenpeace* magazine, September/October, 1989. Reprinted by permission of the author.

"Boundless Bull," by Herman E. Daly, appeared in the Summer 1990 edition of *Gannett Center Journal.* Reprinted by permission of the author and the Gannett Foundation Media Center.

"Overpopulated America," by Wayne H. Davis, appeared in *The New Republic*, January 10, 1970. Reprinted by permission of the author.

"Grass-Roots Groups Are Our Best Hope for Global Prosperity and Ecology," by Alan B. Durning, appeared in the July/August 1989 issue of *Utne Reader.* It is excerpted from a paper prepared by the Worldwatch Institute, "Action at the Grassroots: Fighting Poverty and Environmental Decline." Reprinted by permission of the Worldwatch Institute.

"The Conservation Dilemma" is from *The Arrogance of Humanism* by David Ehrenfeld. Copyright © 1978 by Oxford University Press, Inc. Reprinted by permission.

"Population and Development Misunderstood," by Anne and Paul Ehrlich, appeared in the Summer 1986 edition of *The Amicus Journal.* Reprinted by permission of the authors.

"Waste a Lot, Want a Lot: Our All-Consuming Quest for Style," by Stuart Ewen, is adapted from *All-Consuming Images: The Politics of Style in Contemporary Culture*, by Stuart Ewen. Copyright © 1986 Basic Books, Inc. Reprinted by permission of Basic Books, Inc., Publishers, New York.

From *Confessions of an Eco-Warrior* by Dave Foreman. Copyright © 1991 by Dave Foreman. Reprinted by permission of Harmony Books, a division of Crown Publishers, Inc.

"Apollo's Eye View," by Chellis Glendinning previously appeared in *Woman of Power* magazine, no. 11, Fall 1988, on the theme "Science and Technology." Reprinted by permission of Woman of Power and the author.

"The Earth as a Living Organism," by James E. Lovelock, appeared in *Biodiversity*, copyright © 1988 by the National Academy of Sciences, National Academy Press. Reprinted by permission of the National Academy Press.

"An Iroquois Perspective," by Oren Lyons. In *American Indian Environments: Historical, Legal, Ethical, and Contemporary Perspectives.* Edited by Christopher Vecsey and Robert W. Venables, 171–174. Syracuse, N.Y.: Syracuse University Press, 1980.

"The Future Is Today: For Ecologically Sustainable Forestry," by Chris Maser, appeared in *Trumpeter*, Spring 1990. Reprinted by permission of the author.

"Restoration and Reunion with Nature," by Carolyn Merchant, appeared in the Winter 1986 issue of *Restoration & Management Notes.* Reprinted by permission of The University of Wisconsin Press.

"The End of the Lines," by Norman Myers, is reprinted with permission from *Natural History* magazine, vol. 94, no. 2; copyright the American Museum of Natural History, 1985.

"Do We Really Want Diversity?," by Reed F. Noss, appeared in the *Earth First! Journal* and the *Whole Earth Review*, 1987. Reprinted by permission of *Earth First! Journal* and the author.

"Searching for Common Ground: Ecofeminism and Bioregionalism," by Judith Plant, appeared in the Winter 1987/1988 issue of *The New Catalyst*. Reprinted by permission of the Catalyst Education Society and the author.

"The Cancer Analogy" is from Van Rensselaer Potter's 1988 book *Global Bioethics: Building on the Leopold Legacy*. Reprinted by permission of the Michigan State University Press.

Introduction to *Nature and Madness*, by Paul Shepard, appeared in the book *Nature and Madness*, 1982. Reprinted by permission of Sierra Club Books.

"The Gift of Wilderness," by Wallace Stegner, appeared in 1982 in his book *One Way to Spell Man*. Reprinted by permission of Doubleday, a division of Bantam, Doubleday, Dell Publishing Group, Inc.

"The Current State of Biological Diversity," by E. O. Wilson, appeared in *Biodiversity*, © 1988 by the National Academy of Sciences, National Academy Press. Reprinted by permission of the National Academy Press.

Index

Also Available from Island Press

Ancient Forests of the Pacific Northwest
By Elliott A. Norse

Balancing on the Brink of Extinction: The Endangered Species Act and Lessons for the Future
Edited by Kathryn A. Kohm

Better Trout Habitat: A Guide to Stream Restoration and Management
By Christopher J. Hunter

Beyond 40 Percent: Record-Setting Recycling and Composting Programs
The Institute for Local Self-Reliance

The Challenge of Global Warming
Edited by Dean Edwin Abrahamson

Coastal Alert: Ecosystems, Energy, and Offshore Oil Drilling
By Dwight Holing

The Complete Guide to Environmental Careers
The CEIP Fund

Economics of Protected Areas
By John A. Dixon and Paul B. Sherman

Environmental Agenda for the Future
Edited by Robert Cahn

Environmental Disputes: Community Involvement in Conflict Resolution
By James E. Crowfoot and Julia M. Wondolleck

Forests and Forestry in China: Changing Patterns of Resource Development
By S. D. Richardson

The Global Citizen
By Donella Meadows

Hazardous Waste from Small Quantity Generators
By Seymour I. Schwartz and Wendy B. Pratt

Holistic Resource Management Workbook
By Allan Savory

In Praise of Nature
Edited and with essays by Stephanie Mills

The Living Ocean: Understanding and Protecting Marine Biodiversity
By Boyce Thorne-Miller and John G. Catena

Natural Resources for the 21st Century
Edited by R. Neil Sampson and Dwight Hair

The New York Environment Book
By Eric A. Goldstein and Mark A. Izeman

Overtapped Oasis: Reform or Revolution for Western Water
By Marc Reisner and Sarah Bates

Permaculture: A Practical Guide for a Sustainable Future
By Bill Mollison

Plastics: America's Packaging Dilemma
By Nancy Wolf and Ellen Feldman

The Poisoned Well: New Strategies for Groundwater Protection
Edited by Eric Jorgensen

Race to Save the Tropics: Ecology and Economics for a Sustainable Future
Edited by Robert Goodland

Recycling and Incineration: Evaluating the Choices
By Richard A. Denison and John Ruston

Reforming The Forest Service
By Randal O'Toole

The Rising Tide: Global Warming and World Sea Levels
By Lynne T. Edgerton

Saving the Tropical Forests
By Judith Gradwohl and Russell Greenberg

Trees, Why Do You Wait?
By Richard Critchfield

War on Waste: Can America Win Its Battle With Garbage?
By Louis Blumberg and Robert Gottlieb

Western Water Made Simple
From *High Country News*

Wetland Creation and Restoration: The Status of the Science
Edited by Mary E. Kentula and Jon A. Kusler

Wildlife and Habitats in Managed Landscapes
Edited by Jon E. Rodiek and Eric G. Bolen

For a complete catalog of Island Press publications, please write:
Island Press, Box 7, Covelo, CA 95428, or call: 1–800–828–1302

Island Press Board of Directors